The Ballets of Antony Tudor

THE BALLETS OF

Antony Tudor

Studies in Psyche and Satire

JUDITH CHAZIN-BENNAHUM

New York   Oxford
OXFORD UNIVERSITY PRESS
1994

Oxford University Press

Oxford   New York   Toronto
Delhi   Bombay   Calcutta   Madras   Karachi
Kuala Lumpur   Singapore   Hong Kong   Tokyo
Nairobi   Dar es Salaam   Cape Town
Melbourne   Auckland   Madrid

and associated companies in
Berlin   Ibadan

Copyright © 1994 by Judith Chazin-Bennahum

Published by Oxford University Press, Inc.
200 Madison Avenue, New York, New York 10016

Oxford is a registered trademark of Oxford University Press

Library of Congress Cataloging-in-Publication Data
Chazin-Bennahum, Judith.
The ballets of Antony Tudor :
studies in psyche and satire /
Judith Chazin-Bennahum.
p.   cm.   Includes bibliographical references and index.
ISBN 0-19-507186-7
1. Tudor, Antony, 1909–1987.
2. Choreographers—Great Britain—Biography.
I. Title.   GV1785.T83C53   1994
792.8'2'092—dc20   [B]   93-1088

9 8 7 6 5 4 3 2 1

Printed in the United States of America
on acid-free paper

This book is dedicated to the enduring friendship of Selma Jeanne Cohen and Nancy King Zeckendorf and to my dear father, Maurice Chazin

Preface

Memory and history are far from being synonymous. They are in fact, opposites. Memory is life, it is always carried by living people and therefore it is in permanent evolution. . . . It always belongs to our own time, a living link with the eternal present.
— Pierre Nora, *Les Lieux de mémoire*

In the late spring of 1959, a few dancers were standing at the barre before class with Robert Joffrey. I had been going to Joffrey's studio for some time. He had been one of my teachers at the High School of Performing Arts, and I had also danced in his early performances around New York City. Nancy King Zeckendorf, also a regular in Joffrey's classes, knew that I was looking for a job and encouraged me to audition for the Metropolitan Opera Ballet Company, where she was a soloist; there were a number of places available, and they were going to hold large auditions in the following weeks. She mentioned that Antony Tudor had high hopes for creating an energetic ballet company to perform full ballet evenings at the old Metropolitan Opera House. She cautioned me to show up for class a couple of days before the audition so that Tudor could get a chance to really look at me. I knew only that Tudor had choreographed *Pillar of Fire* and that he had the reputation of being a "strange man." Of course I took Nancy's advice and went to the audition.

Class was grueling, the audition was terrifying, and I did not think I had a chance. There were many dancers who I thought looked very good. The Met's ballet mistress, Mattlyn Gavers, and John Gutman, associate director of the Met, also watched us go through our paces. Tudor bantered and made cracks about why any dancer would want to join the Met Opera Ballet, asking how we could bear to listen to all the shrieking and yowling during those crazy operas. But that was late in the day, after he had chosen those of us who were to enter the company. From that moment on, it has been for me a long, fine adventure with Tudor the teacher, Tudor the choreographer, and his ballets.

Though Tudor's output in this century may seem small compared with that of George Balanchine, Jerome Robbins, or Frederick Ashton,

Tudor's lifetime work is daunting for any dance writer, simply because his artistry gave such amplitude to each and every dance he tackled, whether it was for the ballet stage, opera, musical theater, or television. In his private notes, Tudor, using Henry James's image, mentions that the characters in a ballet can only emerge, like figures in a carpet, after a birthing process that both choreographer and dancers must experience. Starting with his first work in London, *Cross-Garter'd,* created for Marie Rambert's Ballet Club in 1931, he underwent a period of gestation with every one of his ballets, sharing his struggle, his excitement, and his knowledge with the dancers whom he trusted. He kept his private life to himself, but not his making of a ballet. At times he separated himself from his dancers, preferring to shape an idea alone, but it was always his intention to arouse in the dancer the spirit and sensibility to *become* the character in the dance. Now, more than five years after his death, we are able to interpret the body of his work and assess his monumental influence on dancers, choreographers, and the public. We may even say that his full measure can be taken.

One of my most important reasons for choosing to write this book is that, despite Tudor's acknowledged greatness, very little has been published about him and his ballets by dance historians. Consequently, teaching about Tudor in dance history classes became a frustrating exercise. Even worse, his reputation fell into a kind of eccentric limbo where he might be remembered in an offhand way as a postscript to the great choreographers of the twentieth century. As an extreme example of this current attitude, I read about a young American Ballet Theatre dancer in the spring of 1992 who was interested in making dramatic ballets. When asked which choreographers influenced his work, he named George Balanchine, Jerome Robbins, and Merce Cunningham. I bristled at this aspiring choreographer's making no mention of Tudor and realized that he needed a spokesperson not only among dance writers, but also in the working dance world.

I know that others have wanted to write about Tudor in the past fifty years. But just as it was not easy for him to toss out a ballet a season, so it was not easy for Tudor to give anyone the encouragement and access that would have made the formidable task truly do-able in his lifetime. Tudor often said he wanted to be the one to write a book about himself. In the end, it seems he ran out of time or simply chose not to explain himself, finding it difficult to overcome a long habit of personal reticence. At our five or six meetings and conversations from 1985 to 1987, he was receptive and encouraging, perhaps because he liked the idea that I had written a book on eighteenth-century ballet and certainly because I reassured him that this would not be a book about his life but rather a close examination of his work and his artistic contribution to several generations of dancers and choreographers. Thus, this is a book about Tudor's ballets and the way this complex and enigmatic man's career shaped his work.

For years, I have tried to see as many Tudor ballets as were available to study in performance on film or video. For this book, my scholarly focus has been on the premieres of his works, rather than on new productions or restagings of them. As we all know, a ballet that is reconstructed is re-experienced and to some extent transformed. It is very difficult, even with film and notation, to know what original works really looked like, especially overall. No doubt, much more remains to be unearthed about these ballets, and I regret any important gaps, mistakes, or inconsistencies in this research. In the process of describing Tudor's ballets and the work with words, I shied away from post-structuralist epistemologies, from recent linguistic constructs and literary codes. Rather, I sought the gist of Tudor's movement ideas, patterns, and meanings by drawing from my experience as a dancer and choreographer, both technically and emotionally. At the same time, I quoted freely from eyewitness sources that, for good or bad, wielded a significant influence over the reading public at the time of Tudor's ballet premieres. I also interviewed a number of his dancers, finding a wealth of information and insight in the memories of those who danced, studied, and struggled with Tudor. Without exception, they said—in one way or another—that Tudor had changed their lives.

Albuquerque, New Mexico,   J. C-B.
August, 1993

Acknowledgments

My first thanks go to my husband David and my three children, Nina, Rachel, and Aaron, who both tenderly and stalwartly supported me through long research journeys abroad and daily troubles with this manuscript. David provided many photos as remembrances of our trips to visit fascinating people and places and offered numerous suggestions about the book. To my mother, Mary Chazin, I am grateful for her dearly appreciated encouragement, and I shall miss my father and my mother-in-law Midge Bennahum, who enthusiastically helped me in the early stages of this endeavor but, sadly, died before I was able to show them this book.

I could not have initiated this work without the approval of Tudor himself and the several wonderful conversations he permitted me during which he shared significant information about his work. Nor can I ignore the substantial importance of studying for years with Margaret Craske and Alfredo Corvino, who helped me understand the Cecchetti technique that Tudor so valued. Others who generously gave of their time to talk to me or answer my written inquiries include the following: Hugh Laing, Agnes de Mille, Nora Kaye, Herbert Ross, Maude Lloyd, Muriel Monkhouse, Elisabeth Schooling, Fernau Hall (who quoted generously from his biography of Tudor), Alicia Markova, Dallas Bower, Mary Farkas, Nils Åke Haggböm, Gerd Andersson, Viola Aberle, Per Arthur Segerström, Lulli Svedin, Mariane Orlando, Anna Grete Stahl, Anne Borg, Muriel Topaz, Selma Odom (who graciously presented me with a transcription of Tudor's lectures at York University), Martha Hill, Mary Chuddick, Diana Coon, Elizabeth Sawyer, Sallie Wilson, Bruce Marks, Donald Mahler, Sallie Brayley Bliss, Nina Feinberg Brickman, Nancy King Zeckendorf, Isabel Brown, Janet Rowson Davis, Selma Jeanne Cohen, Caroline Bristol Cornish, David Vaughan, Francis Mason, Camille Hardy, George Dorris, Benjamin Harkarvy, Veve Clark, Carl Ostertag, Edward Pask, Jennifer Predock Linnell, Aanya Adler

Friess, Dawn Lille Horwitz, Barbara Horgan, Shirley Eckl Parker, and Tara McBride—and sincere thanks to Marilyn Hunt for permitting me to quote from her interviews, for providing me with so much information early on, and for illuminating suggestions just before publication.

I offer particular thanks for the excellent and notable amount of editorial work accomplished by my good friend Michaela Karni. In the later stages, my sincere gratitude extends to Sheldon Meyer, Barbara Palfy, and Gail Cooper. Kim Carter and Blair Coker provided strong support in the last stages of the book. I am grateful to Claude Conyers for his initial trust in the Tudor project and his encouragement with the manuscript.

I especially thank Linda Bartlett for her diligent and refined work photographing at the Dance Collection, and Monica Moseley of the Dance Collection of the New York Public Library for the Performing Arts at Lincoln Center for her sensitive and sensible suggestions. In addition, I am most grateful to Jane Pritchard of the Rambert Archives in London; Genevieve Oswald, Madeleine Nichols, and Rita Waldron at the Dance Collection; Bergliot Krohn Bucht at the Kungliga Teatern in Sweden; Sarah Woodcock at the Theatre Museum in London; Florence Petten of American Ballet Theatre; Charles Dillingham; Lisa Machlin at the Dance Notation Bureau; Robert Tuggle and John Pennino of the Metropolitan Opera Archives; and J. L. Smith of the Sadler's Wells Theatre in London. All of the above individuals are to be commended for their enduring dedication to the work of Tudor; they also made the task of researching Tudor easier and quite fun.

I must also express my gratitude to the University of New Mexico and the National Endowment for the Humanities for grants enabling me to travel and do research. Portions of this book have previously appeared in somewhat different form, in the Volume 1, part 2, 1989 *Choreography and Dance* in London, and the 1988 *Proceedings of Society of Dance History Scholars* at North Carolina School of the Arts. For permission to reproduce the illustrations, I have to thank the photographers, owners and copyright holders whose names appear in the captions. In addition, I offer sincere appreciation to Damon Von Eiff for his fine work in photo retouching.

And I am beholden to friends and family for being there when needed in the course of this study: Joel and Linda Chazin, Michael Bennahum, Connie Slavin for venturing to Toynbee Hall, Kay Brooks, Stella Fischbach, Judith Berke, Edward Androse, Bennett Simon, Roberta Apfel, Joetta Jercinovic, Jim Linnell, Tom Dodson, Kay Wille, Sandra Edwards, Agnes Harrison, Marcia Siegel, Deborah Jowitt, Elizabeth Zimmer, Christena Schlundt, Elizabeth Aldrich, Eva Encinas, Ginny Wilmerding, Bill Evans, Celia Dale, Mary Mingo, Lucy Hayden, and Barbara Barker.

Contents

Illustrations

The Ballets of Antony Tudor

.

CHAPTER ONE

Introduction to Tudor's Dominions: Studio and Stage

No one but Tudor could have made arabesques and pirouettes say so much or could have mixed such academic artificialities with utterly natural movements to produce a convincing unity.

James Monahan

*A*ntony Tudor changed the way we look at ballet. His manner of teaching and rehearsing as well as his attitude toward dance molded hundreds of dancers and choreographers, prompting them to look closely at and examine the very material of movement. No choreographer before Tudor so used ballet's silent language as a means of understanding characters and the deeper realities of human relationships. He did it by a painfully detailed appraisal of the potential meaning of movements—an appraisal that amounted to genius.

Tudor's choreographic concepts as well as his working style represented a radical departure for the ballet world. His sentiments were close to those of modern dance, wherein a dance's preparation, motivation, and significance of meaning impel the movement ideas. But ballet choreographers even in the 'forties—long after the pioneer dances of Mary Wigman, Martha Graham, Hanya Holm, and Doris Humphrey—worked according to traditional formulas and patterns of steps. As a young man in London, Tudor had studied with a disciple of Mary Wigman's, and was strongly influenced by the powerful dramatic works of Kurt Jooss. Interested in fresh ideas expressed through his ballets, Tudor was to have a vitalizing effect on all kinds of choreographers in England and America.

However painfully achieved, Tudor's vision of dance made an enormous impact on this century, even though many of his ballets are no longer seen. The characters in Tudor's ballets are people, not abstractions; moreover, they are people with a past, whose purpose in a ballet requires more than arabesques and pirouettes. They react to the situations they find themselves in, to other characters onstage, and, primarily, to their own voices.

As a twentieth-century artist, Tudor understood the power of the psyche in motivating gesture and in filtering the complexities of ballet technique.

Various choreographers had experimented with a more emotional ballet technique, but if anyone were to be considered his esthetic parent it would be Michel Fokine, as Tudor often acknowledged, a true pioneer of expressive dance. Fokine wrote a celebrated letter to *The Times* in London on July 6, 1914, explaining the heritage and reformative principles of the Ballets Russes, which was then performing at the Drury Lane. (Although Tudor was a child at the time, he later had the opportunity to see the Ballets Russes.) Fokine's letter begins with the precept, "Every form of dancing is good in so far as it expresses the content or subject with which the dance deals"; Tudor asserted that he needed to discover a movement language for each of his ballets that was keyed to the sentiment of his subject. Fokine decried the use of "ready-made and established dance steps"; Tudor also despised formulaic classroom vocabulary in his ballets. In addition, Fokine wrote that he disliked the use of pantomime or "mimetic gesture" as a diversion to amuse the spectator. Rather, he sought to use the whole body, not just the hands, to further the meaning of his scenario. Similarly, Tudor created poetic gestural moments without melodrama or histrionics. According to Fokine, ornamental ensemble dancing must also be eliminated. If the choreographer arranges groups of people on stage, they must be connected to the action and respond to one another in an expressive manner. Indeed, Tudor rarely engaged the corps de ballet in long series of unison work. These precepts represented a new step in the development of ballet choreography, taking the ideas of Michel Fokine farther and assimilating them with the theories of twentieth-century modernists—dancers, writers, and painters.

Certainly there were other brilliant ballet talents of the post–World War I period, choreographers such as Bronislava Nijinska, Kasyan Goleizovsky, Léonide Massine, George Balanchine, Kurt Jooss, and Frederick Ashton. With Tudor, the ballet world had to stop and take notice. His work seemed to demand more of the viewer, not because the movements differed so much from those of previous choreographers, but because they dealt with the private thoughts of very real and very serious people.

While still a young dancer, Tudor often attended the theater, where he saw works by Henrik Ibsen, Ivan Turgenev, and T. S. Eliot, as well as many of Anton Chekov's plays. In fact, English critics came to call Tudor the "Chekov of the ballet." Tudor tended to emphasize character rather than plot and to create people with a past, a present, and a future, using a movement vocabulary suited to the sympathies and reflections of his characters. He could conjure up a human personality, a social setting, or a complex situation with one or two deft strokes. He was less interested in surface action than in the emotional history of an event. His ballet scenarios are filled with characters quite different from the usual ballet personages, and he scorned the "frou frou" and acrobatics he associated with Massine and the later Ballet Russe companies. Tudor's people exude an intense emotive power, drawn in carefully detailed emotional outlines.

He achieved that intensity of emotion on stage not by wallowing in feeling but by holding back. He explained, "My approach is very dry because I always want to know how it was done, why a particular effect has been created. . . . I try to explore the relations of men and women to one another, but I don't jump into anything. I would say that the predominant characteristic of my choreographies is control, control of physical movements, of main effects. Nothing in my ballets is ever fully stated."[1]

All the roles in *Gala Performance* and *Judgment of Paris,* Caroline in *Jardin aux Lilas,* and Hagar in *Pillar of Fire,* as well as the male protagonists, such as the Transgressor in *Undertow,* Romeo, and the Boy with Matted Hair in *Shadowplay,* demand intense, intelligent, and relatively restrained interpretations. Caroline is a thousand women in each generation; she is every wounded woman in every story of forced marriage; nor are ballerinas in *Gala Performance* specific ballerinas; they are the French, the Italian, and the Russian. Because Tudor was able to say so much about each of these people in simple, expressive, and precise movements, he did not overpaint— although the dancers often do. Rather, he suggested certain qualities, such as the sorrowful lyricism of the Lover in *Jardin aux Lilas,* or the cynicism of the hip-swinging decadent women in *Judgment of Paris.*

For this purpose, the dancer's analytical ability came into play. Tudor declared that "the performer should seek to achieve economy, but with perfection in purity of technique and manner." He focused on the dramatic motivation of a dancer playing a character and looked for distinguishing and characteristic qualities. He disclosed that "manner becomes the power of producing emotional response in the beholder." For him it is the indispensable element of theater. "If of theatre, why not then of ballet and of classical ballet? Perfection of manner is assuredly more rare in classical ballet than the acrobatics that have so often passed for the whole when they are but a part. In the achievement of this manner lies the ability to create atmosphere and thereby to gain the response of the audience."[2]

Nora Kaye, one of the great dramatic ballerinas of our century, often spoke of her experiences with Tudor and of his desire to remake her into each of the personalities that she danced. As Hagar in *Pillar of Fire,* she and the ballet could not fail, he felt, even if the steps were not executed exactly the same each time, because she would *be* Hagar and therefore what hap-

5

pened to her would be believed by the audience. Kaye remarked that taking Tudor's ballet classes helped her to develop her character onstage and that the flow from class to rehearsal to stage, manipulated and dominated by Tudor, became a natural continuum, enhancing her understanding of Tudor's work. But the painstaking process of learning how to become a character in Tudor's dances often took an enormous toll of both the dancers and Tudor.

In preparing his dancers to play their parts, Tudor understood the nature of the dancer's life, how enclosed and isolated the experience of living in a studio can be. At the same time, he saw the studio as a laboratory, an experimental space where it was safe to examine human relations in a dance context. The unique qualities of Tudor's characters could be realized only by ballet dancers willing to learn about themselves and to shed their dependent mannerisms. As a result, he insistently pressed his dancers toward self-examination and self-awareness. This spirit of Socratic inquiry underlined his rehearsals as well as his classes: "Did you ever think about giving up who you think you're supposed to be?" "What did you just do?" "Why did you do that?" "What were you thinking as you accomplished this step?" "How did you feel as you were doing that movement?" "Where were your arms when you were making that movement?"[3] Tudor taught that artistry was predicated on responsibility, on the search for meaning and truth, a preoccupation that in his later years would make him a Zen master.

Tudor needed quick-witted and imaginative dancers in order to achieve his rich but understated movement vocabulary. Therefore, he chose extremely musical and malleable people, with emotional breadth and warmth. He liked his dancers to be technically proficient, enough so that classically fine technique looked nonexistent. Since the structure of his ballet steps was built on an emotional tension, the technique might therefore be twisted or disguised. In order to accomplish his purpose, he insisted that the musicality of the dancers must exceed their dependence on technique, so that the movement phrase was never broken.

Tudor initiated his ballets ritualistically. He developed a distinct language of ballet movement for each of them, focusing on signature or key movements that subtly evoked the important psychological qualities of his protagonists:[4]

> You begin by devising movements to establish the language in which you will work, movements conveying the spirit of the ballet. . . . I go knowing my musical score thoroughly and I have the dancers, the right ones hopefully . . . dancers must be trained; that's the acting ability. That's why it's theatre and not the street. In addition the same movement can mean so many

different things, which is partly timing, again proper breathing, partly the musical background.[5]

He described first rehearsals: "In the beginning, I'm rather dilatory. I don't know what I'm going to do until I start rehearsing, which drives the dancers mad. From the sensations in the body, you do your own choreography. There's no way else to do it. You feel your own muscles. The only rewarding way of working is for the choreographer just to give himself absolute freedom and let the body go and let happen what may."[6] He called himself "uncanny. I never know how those things happen. If it happens, it happens. If it doesn't, I wait a little longer and then it does."[7]

In a lecture given on April 8, 1951, at the Henry Street Playhouse in New York, Tudor contemplated the state of ballet and the difficulty of being taken seriously by an audience too used to screaming "bravos" for virtuosic double *tours en l'air:* "So the modern ballet choreographer has to keep ahead in the hope that, though he may not catch on today, he will eventually. We ballet choreographers have to examine ourselves and our material much more thoroughly than we do. I don't think we explore our movements nearly enough. We seem to feel that the ballet has been examined enough for centuries, so we have little desire to look into it further." He decried the fact that modern ballet could not stand by itself the way music and literature could: "Dance should develop from dance itself and a more conscious knowledge of the use of dance. The choreographer has to explain by teaching, by opening hearts and minds, not by telling people how to do steps. There should be a constant opening of doors by the teacher for the student. . . . Just as you can't expect the best of a musician in the pit without a good conductor, so the choreographer should be able to draw the best out of the dancer."[8]

During rehearsals, Tudor's terse speech resembled his choreography—carefully chosen, metaphorical, asking the dancer to fill in the empty spaces. Despite his reputation as the "father of psychological ballet," his motivating concern in the making of a ballet was not psychology: "It worked itself into the stories of *Pillar of Fire* and *Undertow.* The ideas came of themselves." Tudor compared himself to a "medium through which ideas flow in rapid succession. My job is to select and arrange them. If a dancer doesn't catch a movement as it flashes fresh from the creative fire, there is no catching it at all. . . . There is an incubation period during which time I go around in a trance planning the new ballet. But often movements that seem wonderful in my imagination don't work out well and I need to change them."[9]

Tudor studied his themes thoroughly and brought his dancers into this process, occasionally through readings, but also through suggestive and improvisational methods. Provoking the dancers' imagination became part of the rehearsal process. As Janet Reed remembered, "Tudor used wonderful imagery to help the dancer create a role. For the kittenish younger sister in *Pillar of Fire,* he told me that she never looked at anyone directly; it was always from down and under. She never went straight to anything; her movements were all curly. In *Jardin aux Lilas,* he told me to move as if I were stepping on grass; in *Gala Performance* it was a red velvet carpet. He never worked on character as a thing apart from dance; the character was in the movement."[10] Steps had to have a motivation, a purpose; and the long hours of rehearsal tended to ensure that the dancer caught on to Tudor's character sketches. The notator for the ballet *Undertow* said, "I remember sitting next to him while working on *Undertow* hour after hour rehearsing with him. We would review and labor over each character and then move on. Tudor's body, and especially his back and neck, would change shape and quality as he watched and rehearsed the various characters. He might not remember the exact steps. He kept fussing. He would do fourteen versions and then come back to the first version. He wasn't happy with what he did if it didn't match his mind's eye vision."[11]

While the dancers struggled to "become" the characters, Tudor wrestled with one of the most difficult issues in choreography, the music. He acknowledged this concern to Zita Allen, in the *SoHo Weekly News* of May 3, 1979:

> Most ballets are carried by their music: I take a long time. I look for the music, search for it, shape it. In order to shape the score I know roughly what the ballet's going to be. I know who the people are, but I don't know what they're going to do. Let's face it, most ballets are carried by their music and not their choreography.

Of course, it might seem that the musical score served also as the ballet's script. Not so:

> I try to avoid doing too much with the music. I think it should be a companionship. There are certain composers who really do not lend themselves to choreography very well because the music says it all with such force and such completeness and gives no malleability to the work. If the music dictates too much you have to get rid of it.

But also, much depended on the way the choreographer asked the dancer to execute a combination of movements musically. At the height of *Jardin aux Lilas*'s emotional tension, the moment when all motion onstage freezes in the so-called "sleepwalking" scene,

> Caroline has a series of simple, unadorned walks, which take her from one frozen grouping of dancers to the next. The pattern is then traced to take her

back in time, space, and in the drama to where she began. The walks are one or occasionally two to each musical beat. The dancer may neither add nor subtract any steps; there must be the exact number or the musicality is lost. In addition, their direction—some forward, some sideward, diagonal and backward, what Tudor calls the geography—is absolutely fixed because of the placement on stage of the various groups she must visit. . . . Whether the dancer has long legs or short, those inexorable walks may not be altered, nor may her destination be changed by the size of the steps. She must perform the material exactly as it is set, with no leeway, or the whole drama is lost.[12]

When asked to articulate how the music inspired him to make movement, Tudor responded:

I simply don't think of the rhythm of dance on a beat-for-beat basis. I think of the phrase and the line and if the music breathes, the dance can breathe with it. And if the dance does not coincide with the music at every point, I am not concerned. There is a danger in using music that makes the listener dance with it. In that case the dancers might just as well not be there. Such strongly rhythmed music is useful only to cover up the lack of choreographic ideas.[13]

Tudor did not manipulate "the most obvious melodic or rhythmic line. Occasionally he would build a phrase that had seven counts with a six-count phrase. Or he would repeat the phrase with movements to five counts leaving out a count; he loved jagged rhythms. He rarely used a straight $\frac{4}{4}$."[14]

Dancers' counting has to be about phrasing and performance. Tudor felt as passionately as anything that you could not chop up performance music, that you had to find a way to go through it. "He did not want to give the dancers any kind of mincemeat." He would ask, "Where are we supposed to be on count 4? If you have to count it, then it's a lost cause." For him, any kind of musical experience was physical: "How do you dance within a phrase, how do you get the kinetic thread of the movement which creates the expressivity? Most importantly, how do you not sacrifice the dance for the steps while still keeping the feeling and dynamics going?"[15]

It was not always pleasant for Tudor's dancers when he pushed them in directions they had never explored. Not many could take the strain of his cruel cajoling and intransigence, although he tended to be more lenient with students than with professional dancers. When asked how he wanted to be treated by his dancers during rehearsal, he quickly responded, "Like God!"[16]

But he was harder on himself than on his dancers. One of the qualities that most distinguished Tudor from his contemporaries was his uncompromising view of his own work. He always knew what looked right for his ballets, but the agony he suffered in order to achieve the clarity and flow of a scene and the musical and dramatic accents of each movement would often overwhelm him. At times, it was not until after the premiere that he finished making the ballet. The openings of *Dark Elegies* and *Romeo and Juliet* were actually delayed, and forty-eight hours before the opening of *Shadow of the Wind* there still remained many gaps in the sixty minutes of choreography. This exasperated both the dancers and the producers, but he would not present to the public what had not fully made the transition from his own mind to the stage. He was often onstage with groups of dancers just before the curtain rose, rearranging different combinations of movements. Terror would rise in the heart of the dancers at the prospect of remembering these last-minute changes.

As Nora Kaye said, the flow from class to rehearsal to performance shaped the choreographic process; a long journey for the dancer that usually began in Tudor's classes. When entering a company class, one had to sense what Tudor's mood would be; the positive energy of the group could assuage his ever-trenchant critiques. Yet, no matter what his humor, he always taught an interesting class. For example, for center work he might play with movement ideas that popped up later in rehearsals for his ballets. Occasionally, he chose a theme that he carried through the entire ninety-minute class. He might work on the concept of the *grand battement,* a sweeping movement of the leg up to its highest point and then back to the ground. *Grands battements* commonly exist in all of Tudor's ballets. The energy of the leg is started in the foot, from the moment the lift is initiated. Done correctly, jumps become much higher and suspension in the air longer, while executing *sauts de basque* and *grands jetés* or *tours jetés.* Consequently, his classes revealed some rather simple ideas about technique that had not been thought of before. Tudor never did anything that was easy, always searching for the new, pushing, going beyond the obvious. His suggestive imagery often diagnosed a technical problem: "Do this first *port de bras* as if you just got up and it's raining."

Ballet class presented issues the dancer would be faced with in rehearsal the following day. "His classes were like choreography," his accompanist reflected. "The combinations were quite complex and he'd only show them once. Rhythmically they were complicated phrases, based on the premise that you could not approach them as steps. They were always connected with an underlying thread and overlying shape, as with *glissade jeté, glissade jeté, jeté, jeté, jeté assemblé.*" But Tudor usually created fascinating combinations of rhythms to build an amusing and challenging phrase. "For example, *glissade jeté, glissade jeté, jeté, jeté,* hold, hold, then start the next phrase with *jeté assemblé,* which would finish off in the middle of a phrase. Furthermore, Tudor required his students to sing the music. That would open the chest and the throat and would help them to learn the flow of the phrase. The other device he would use would be to create an exercise in $\frac{4}{4}$ or $\frac{2}{4}$, and ask to have it performed in $\frac{6}{8}$ or $\frac{3}{4}$."[17]

Tudor's classroom dicta remain important to those who had an opportunity to hear the full force of his ideas. Movement for him and for his students had a way of being extremely sensory, even sensual. For example, two positions, such as *croisé* and *effacé*, differed not only in the lines they formed, but also in the way one's body felt doing them and, in the end, what response they evoked in the viewer. Dancers were always aware that what they worked on in class was part of a search for values that would and should affect the performance on the stage.[18] Tudor sought to arrive at principles that explained where movements initiated. He stated many times, "Center-of-the-body consciousness is a primary necessity that far outweighs the mechanics of high kicks, etc. . . . an unmalleable spine, with automated legs and flappy wrists can only produce the non-genuine."[19] Using the central weight of the body or moving from the center of gravity helped the ballet dancer focus on where a movement starts. He encouraged dancers to move from the torso, naturally taking the hips along. The arms as well had to have a fluidity and prior motivation. This awareness tended to relax the limbs. Like August Bournonville, the nineteenth-century Danish choreographer, Tudor abhorred preparations or showing that "now I shall do something important." He believed in coherent work, where the dance phrase, the music, and the emotive qualities all come together, and no one step is featured. Hugh Laing explained:

> Tudor may expect the dancer to do four pirouettes, but one cannot let the preparation show. If the turns are part of a phrase that may be saying, "I love you Juliet," you must not interrupt that phrase to take a fourth position preparation because then you are paying attention to yourself as a dancer, not to Juliet.[20]

More than any value, "Tudor cared about spontaneity, an almost improvisational approach, not the studied, dry, academic outlook. A certain amount of risk-taking and daring as well as exposing one's inner self brought excitement and life to one's dancing."[21]

It is important to remember that Tudor studied ballet in London with Margaret Craske, a teacher whose passion for the Cecchetti ballet technique amounted to religious belief. With Tudor, professional class involved a daily inquiry into variations on the Cecchetti style of moving, a dance method that differs considerably from the Russian syllabus, and that the Royal Ballet in England perpetuated for many years. Enrico Cecchetti (1850–1928), the famous Italian ballet master and private tutor to Anna Pavlova, taught

Tudor's major teachers, Marie Rambert and Margaret Craske. From the Cecchetti technique, Tudor learned an evolutionary dance process that developed the gradual tuning of the body's movement in space. Cecchetti's explicit placement of the body, the phrasing of his exercises, and the careful and exact use of arm movements or *port de bras* helped to form strong as well as lyrical dancers. All of Tudor's choreography is fundamentally based on this technique, so that a familiarity and understanding of it helps solve technical problems in Tudor's ballets.[22]

Although Margaret Craske, his colleague at the Metropolitan Opera Ballet as well as in London, proved an excellent pointe teacher for the ballet company, Tudor demanded a particular way of using the pointe shoe and explained how it should work in his ballets. He did not want to see how one struggled up onto or down from pointe. In essence, he did not want to know that the dancer was wearing pointe shoes. Consequently, Tudor's choice of steps shied away from showing bravura or virtuosic *relevés* or hops on pointe or multiple turns in a circle, etc.

Tudor's idiosyncratic use of the ballet vocabulary in his ballets took time to understand. Many years ago, the critic Rayner Heppenstall, when he first observed Tudor's ballets in London, noticed that Tudor gave particular emphasis to the torso and face: "The most obvious general stylistic achievement, in what Mr. Tudor has so far produced, has been to give the upper part of the body more significance, a wider, more complex range of plastic importance, than it has previously had."[23] He often spoke about the mind's eye, having eyes in the back of one's head, and having a lively back and a long and lithe neck. Dancers were asked to sense what they looked like without using the mirror, and to relate to other people onstage without necessarily seeing them. The back had eyes; the chest had eyes. Above all, one could be entirely focused without actually looking at any one thing.

Not only did the torso play a significant role in Tudor's work, but his use of the arms also surprised dancers and spectators alike. He was not interested in the visual design of a particular ballet movement. Rather, he employed the arms functionally. He did not want to see them unless they had a gestural meaning so that they would not distract from the sense or the flow of the movement.[24] Many of his ballets play down balletic arm positions, although the arms certainly move. One finds the arms resting by the body, taking particular shapes as if punctuating the dancer's thoughts. In *Jardin aux Lilas,* the women consistently keep the arms softly relaxed; they hang by the side of the body, mirroring the long torso and long dresses and giving the body a lengthy, Aubrey Beardsley look.

Always searching for the perfect arm and hand gesture onstage, Tudor pared down movement to the uncluttered essential, yet in class he cared very much about design and insisted on the classical arm positions. Tudor's sensitive use, or non-use, of the dancer's arms and hands in his ballets became a central issue in his technique classes. He spoke of the arms as extensions of the torso and as silent voices of the music rather than convenient, automatic shapes we learn as youngsters in ballet class. This unorthodox arm play seemed a perfectly suited accompaniment to the seamless quality of his complex foot patterning.

Tudor developed exercises to sensitize dancers to the intention as well as the physical execution of dance movements. During one of these exercises, he asked the dancers to stand and move their arms while imagining different-colored lights playing on the torso. How would they express a blue or red or yellow light? Another exercise, almost Zen-like in its focus, involved a sitting dancer who pretended to be an umbrella. The dancer imagines that the umbrella first opens fully, then closes, and finally wraps around with the button clasped. These sensations happen without moving one muscle. The rapt class watched closely and decided, along with Tudor, whether the dancer succeeded in evoking the internalized process. Although these were exercises that he used primarily in his dance production classes at Juilliard, he often stopped rehearsals with a ballet company to interject them in order to drive home a point.

Tudor's rhetorical style of coaching also had an impact. A well-known story pertains to a rehearsal of *Jardin aux Lilas* when Tudor spoke to the cadet, who had just kissed the hand of the fiancé's former lover. The cadet, standing with his weight slung into his hip and one shoulder forward, was asked by an irritated Tudor what kind of costume he was going to wear in the ballet. The young dancer responded, "A uniform." Tudor continued, "What does it look like?" The dancer said, "It has silver or gold braid on the jacket and epaulettes." "Right," Tudor said and directed him to stand with his hips facing the wall. "Now, say the words 'Sergeant, Sir.'" Suddenly, the dancer realized he was a person in uniform and stood erect, his shoulders back over his hips and all his preciousness erased. Tudor yelled, "Don't you ever try to be affected again!"[25]

The dancer who focused on himself rather than the choreography became anathema for Tudor. He declared:

> The visual beauty of dance involves the sequences of rhythms in phrases of movement, and dancers who have primarily supposed beauty to mean that they should look beautiful at each and every moment have unwittingly founded the glorification of the human body cult of second-rate dancers.[26]

Tudor disliked dancers who showed off and thus lost the dance they were doing:

I would like to tell all dancers to forget themselves and the desire for self-display. They must become completely absorbed in the dance. Even in a classical variation there should never be any thought of a dancer doing a variation—he should become identified with it.[27]

Tudor became irritable with dancers who sought explanations of their roles, and in a less generous moment he testily asserted:

I prefer just to shake a shoulder or an arm or a head. Sometimes a dancer I'm working with will say, "Who am I?" And I tell them, "If you don't know by now, either my movements are wrong, or you're not understanding them."[28]

He stated that he would not communicate literal information about his ballets, especially through the mind, to his dancers, "considering what some dancers' minds are like. They must grow into their roles through the movements. Otherwise it's only veneer."[29] But whether he wanted them to delve into the depth of their souls or not, the result of the alleged rehearsal trauma was riveting. A reviewer commented:

Tudor's dancers become their roles. One comes onstage and you're certain that a life for Hagar or Caroline preceded this glimpse you're getting and will continue after she's gone from sight. Dancers act, not just through cleverly manipulated facial expressions, variations on theater masks or traditional pantomime, but with their entire bodies.[30]

Furthermore, Tudor confided:

I can't stand dancers who love to dance. . . . Ninety-nine dancers out of one hundred think that the theatre was created for them. That's wicked. Only one in a hundred realizes it is for the audience. One of the best performances of *Pillar of Fire* we ever gave was the last time I danced in it. John Kriza was going to take over my part and I said to the other dancers, "Johnny will be watching in the wings, and I want you to give a cold-blooded performance, exactly like in the classroom—be precisely on the beat and make your movements letter perfect." People came to us backstage afterwards in tears and said it was the most moving performance they'd ever seen. The power should be in the choreography. It should speak for itself.[31]

Tudor even lambasted some dancers here in America who had become national treasures:

I dislike stars if those stars believe in their own stardom, if they live their lives believing the legend that's been built up around them. I hate them, and won't want to use them in my ballets. I think being onstage is a golden opportunity to escape from their own miserable selves. Besides, stardom comes and goes, doesn't it?[32]

Central to Tudor's esthetic was the struggle for the truth that both the choreographer and the dancer bring to a ballet. Thus, the teaching process became one with the rehearsal process, and no matter what the scenario, the choreographer guided the dancers as they lived through their steps and their role.

According to Dylan Thomas, poetry is "the rhythmic, inevitably narrative movement from an overclothed blindness to a naked vision." Tudor's ballets are often called "poetic," evoking images and powerful personal feelings that resonate throughout his work. Occasionally the so-called poetry in his ballets became maudlin, overacted, and sentimentalized, as in a moment in *Echoing of Trumpets* when several women fight over a crust of bread. But in the main Tudor knew the value of understatement; one word, one gesture, one authentic and essential "vision." Moments of extraordinary clarity vivify almost all his ballets. They reside in his unfailing sense of the factual and psychic rhythm of events that identify his characters. As John Percival has remarked, "His is a poetic realism, not 'poetic' in the sloppy would-be atmospheric, vaguely drifting sense, but in its real meaning, with images that stand firm and solid and ring out bright and clear."[33] In *Jardin aux Lilas,* when the Episode in His Past reaches for her face with one hand, fingers outstretched, powdering, primping, she clearly tries to conceal her narcissistic self as one of the guests kisses her hand. The angry male in the *Dark Elegies* solo taps his foot against the floor in a gestural tic that betrays his demonic fury. The love-hungry Aristocrat's secretive rotating of her uncontrollable hands in *Gallant Assembly,* however humorous, reflects the hidden habits and fantasies of a woman with overwhelming appetites. The woman in the finale of one of the rapturous duets in *The Leaves Are Fading* offers herself to her lover as he lifts her onto himself, her knees bent, her torso arching languorously back as he turns her around and around in dizzying twirls.

Another quality intrinsic to the poetic ambiance of Tudor's work is the play of time and rhythm in his processing of events. As R. P. Blackmur maintained, "Poetry names and arranges, and thus arrests and transfixes its subject in a form which has a life of its own forever separate but springing from the life which confronts it."[34] Tudor often suspended or shaped time by penetrating and exposing the inner impulses of his characters. The central moments in *Pillar of Fire* find Hagar caught up in her tumultuous feeling of rejection by the Suitor and sensual hunger for an habitué of the house of ill repute across the way. After the disturbing *pas de deux* seduction, she exits with him, her legs hanging on his chest in an unmistakably suggestive image; yet one moment later, she re-enters the stage and kneels repentantly. A lifetime has passed during that instant, and in a rapidly crescendoing swirl of events, Tudor has convinced us of all that has happened, leaving us to deal with a sense of her shame. In *Dim Lustre,* Tudor encapsulates the illogical flow of memories suggested by a whiff of perfume or the waft of a hand-kerchief. Months perhaps years, of souvenirs of love are compressed into seconds by using the simple device of presenting his lovers with mirror images of themselves.

Tudor's manipulation of the musical score in his ballets helped enhance their metaphorical quality. For example, in *Jardin aux Lilas,* this "poetry" comes alive when the lush romantic melodies of Chausson are repeated, sounding quite different for each dramatic moment, depending on Tudor's choreography. Stephanie Jordan has pointed out:

> One of these tunes, the main theme of *Poème,* occurs unaccompanied with the first entry of the solo violin. Isolated in this way, it becomes particularly significant, and Tudor has it coincide with an equally significant dramatic moment, the first opportunity for Caroline and her lover to dance together.[35]

During this instant, we, the viewers, are clandestine partners in the couple's compelling restraint of feelings. Each time the motif appears, we are drawn closer to the story's impassioned theme.

Revealing personal truths is the fascination inherent in Tudor's more stirring characters. These people, however, need not be the tragic figures of his serious works. Tudor was extremely proud of his humorous ballets and knew that he had a knack for the English tendency to witticism. As in real life, when Tudor, with a clever verbal thrust, struck out at his dancers for their solipsisms, so he discovered the inherent weaknesses in his characters with a quick brushstroke. *Fandango,* a slight work set to a driving baroque score by Antonio Soler, contains some wonderful moments of parody, such as when the dancers unexpectedly begin to sing out as flamenco artists do, almost screaming as they compete to upstage one another. The three *Gala Performance* ballerinas have personal idiosyncrasies that are painfully symp-tomatic of all female ballet soloists. Whether French, Russian, or Italian, our prima ballerinas might certainly find themselves in their company. The ballerinas' treatment of their cavaliers remains one of the nastiest and most insightful satires on the *pas de deux.*

But Tudor's most salient comic moments occur in *Judgment of Paris.* The buxom Venus executes a double hoop dance as she waves the hoops alternately in the air and passes them over her leg in an attempted double *rond de jambe.* Her tour de force consists in raising both hoops over her head and pulling them down over her body to the ground, a feat that she seems particularly relieved to have accomplished, given her ample figure and the enormous pink bow on her derrière.

Generally, the stories in Tudor's more serious works revolve around individuals struggling with vicissitudes and difficult events. One of the most pervasive of his themes, Freudian in inspiration, emphasizes the notion that sex is power and that the person who controls sex has the power. Many of his ballets focus on one character's hypnotic, seductive authority over another, symbolizing perhaps Tudor's harsh, often brutal attitude toward human weakness. His people can be manipulated, pushed, enslaved, and demeaned by their animal drives.

In *Shadowplay,* for example, the Boy clearly experiences an arousal of his sensuality and, briefly, seems overwhelmed by both the Terrestrial and the Celestial figures, but in the end he comes to find certainty in his own identity and tranquillity. In *Pillar of Fire,* Tudor distances Hagar from her counterpart in the Dehmel poem by throwing her into a frenzied conflict with herself and her sensual appetites. Her *pas de deux* with her seducer remains one of the finest theatrical examples of sexual seduction and conquest, with its violent, painful consequences. *Judgment of Paris* presents a more cynical, sardonic view of sex. Tudor's three jaded goddesses, defeated and down and out, manage a Pyrrhic victory over the gentleman drunk, who almost stays awake to observe their dragged-out, tawdry, sexpot routines. In *Undertow,* the mentally ill and dispirited hero, The Transgressor, kills a prostitute whom he attempts sex with and who obviously symbolizes his mother. From the humorous *Gallant Assembly* and *Knight Errant* to the serious *Cereus* and *Tiller in the Fields,* Tudor's depictions of erotic behavior have "no vulgarity in their obscene images";[36] rather, they breathe an air of sly criticism as well as tolerance into his penetrating characterizations.

Many of Tudor's ballets are dominated by romantic love themes that portray the play of passions, the tendernesses, and the subtle depths of love's experiences. They often reveal mysterious and ardent secrets that thread their way through his scenarios and that perhaps emanate from the sexually repressed world Tudor grew up in. Agnes de Mille has asserted that "in spite of Tudor's homosexual preference, he created the most brilliant, tender and moving pas de deux. He understood love, he put it on the stage again and again and again."[37] The protagonists of Tudor's ballets are conditioned by the complex situations they find themselves in, and the situations vary considerably from one to another. Romantic love carries a heavy price in Tudor's hothouse stories, reflecting his tendencies to Darwinian thinking; he knows the power of context and environment. *Jardin aux Lilas* is a ballet about control and social pressures, not just an Edwardian drama; sadly, Caroline suffers imprisonment in the same way that any of us may be entrapped. In *Dark Elegies,* villagers patiently sharing their grief provide one another with stamina and a sense of purpose. Their love helps them surmount inexorable misery. One of the later works, *Sunflowers,* recounts the tale of four young women who are willing, even eager, to give up their friendship for the sake of love. When the men leave, the women are alone, facing in different directions. Quietly, almost imperceptibly, Tudor presents the truths we live with every day and the solitude of the individual struggling with encroaching experiences.

One question that stands out as particularly pertinent today concerns Tudor's attitude toward women. This does not allude to his treatment of women in the rehearsal studio—which was often cruel but perhaps no crueler than to men—but rather asks if Tudor's dramatic characterizations rest upon a preconceived and prejudiced dislike and dishonoring of the female gender and if his scenarios hint at or display misogyny. In many ways, Tudor reflects the sensibilities of his time. In early twentieth-century European literature, a woman was inevitably portrayed as a creature of her feelings, driven by love and emotion, tempered and perhaps confused by her newly understood "inalienable rights." In most cases she had to be watched over, taken care of, succored, and caged. Often, she dragged men along with her to their lowest depths. Authors whom Tudor admired, such as Anton Chekov, D. H. Lawrence, Henrik Ibsen, and Federico Garcia Lorca, presented women out of step with their worlds and on their way to a fall. These heroines may have sought freedom to be whole, complete, and able to express their feelings without losing face or status, but they were usually condemned for daring to seek that freedom. Their overwhelming passions cast them into an abyss.

Although Tudor was influenced by this literary tradition, it is more likely that he identified with his protagonists, especially Caroline and Hagar, who lived in restrictive and stilted social environments. Since these characters are undoubtedly aspects or projections of Tudor's self, one feels these analytical and harsh perceptions are not directed against women as a gender. Rather, Tudor examined the aspects of human nature that he sought to understand. He appreciated the complexity of the problems his protagonists confronted, which involved sex and love, nobility and baseness of character.

When one looks at Tudor's humorous work for his position on women, however, one remarks on his inclination to highlight the seamier, more selfish and competitive side of women's behavior. In *Judgment of Paris, Gala Performance, Gallant Assembly,* and *Knight Errant,* men are duped and manipulated by scheming ladies with powerful appetites or commanding authority. As society's victims, these women resort to devices that place them in unattractive postures.

Especially offensive qualities are pervasive among the women in *Undertow,* who cheat, seduce, and drive others to murder. Some scenarios, such as that of *Tiller in the Fields,* intimate that women lead men down the primrose path whence they shall never return. The women of Tudor's ballets, however, are too complex, too interesting, and too compelling to be victims of misogyny.

In Tudor's dramatic ballets, he masterfully paints the complex layers of both his men and his women, and in several serious pieces—*Dark Elegies, Romeo and Juilet, Jardin aux Lilas, Shadowplay,* and *Undertow*—he faultlessly depicts the pain, guilt, and suffering of his male leads. But does he reveal the truth about his other male characters with the same shimmering intensity and responsibility to detail and visible clarity? Tudor's ballets occasionally portray men as stereotypes. Remarkably, when Tudor or his lead dancer Hugh Laing danced these generalized roles, the characters became rich and interesting people as if by some magical transformation.

The following chapters examine Tudor's early ballets and the development of his choreographic style. Although his growth is apparent from the very beginning, Tudor's vision of ballet remains constant. His dedication to the scenario, to the ballet vocabulary, to gestural details, and to his particular sensitivity to the music as a partner in the dance nourished his dancers and brought them to an understanding of ballet as a complex matrix calling upon a personal commitment to ballet's many sources and voices.

CHAPTER TWO

Early Years: Childhood in London and the Discovery of a Passion

*A*ntony Tudor was born William John Cook on April 4, 1908, and lived on Central Street in Finsbury. It was a tough, working-class neighborhood in the East End of London, and Tudor learned a good deal about the limitations of his background as he traveled to the local school and eventually found work in the Smithfield Market, where his father was a butcher. Tudor described his father, Alfred Robert Cook, as not very tall, five feet eight inches, extremely strong, used to chopping meat, and an occasional violin player. He enjoyed taking his son to music-hall shows. Tudor's mother, Florence Anne Summers Cook, played the piano when she was young, and sang on Sunday evenings and at parties. "No self-respecting English person would be without a piano in the house."[1] She gave him his beginning lessons, and he continued to study piano until he was eighteen years old.

As a boy, Tudor also participated in the church choir: "I thought I was a good soprano, but I never got to solo." He quit when his voice changed. As a shy and sensitive child, he recalled being very frightened by the bombs going off during World War I. At school, he enjoyed arithmetic, English, and geometry. "I had a wonderful sense of shape which is useful to a choreographer." He often escaped to the "lab" at his council school, where he enjoyed drawing flowers and natural objects. "I was a perfect student, of average intelligence, but I never had teachers who took me under their wing. I remember we read many books and I enjoyed having to make a report from a work on the history of costume."[2]

When Tudor graduated from the free (government-sponsored) school at fourteen, he received a scholarship to a private academy, Dame Alice Owens' School, where he stayed for two years. There he perfected his talent for languages and learned to do accounting, a skill that he put to good use throughout his life. (In later years, Tudor always filled out his own income tax forms, an activity that daunts most of us.)

One senses that Tudor was a child of deep solitude and introspection. "I'm still solitary. I had a brother; he had his own room and I had my own room. The rooms were small, but our own." Tudor mentioned that he was not close to his older brother, Bob, who was very different—more athletic and competitive. In spite of this, in later years they always wrote to each other. His brother lived his whole life outdoors and eventually entered the forestry service and worked in New Zealand, where he and his wife settled with their three children.

As a teenager with a strong, sensitive imagination, Tudor read philosophers, such as Arthur Schopenhauer, and books about religion. His parents took him to several London music halls and theaters. Aside from the wonderful skits of the Christmas Pantomimes, according to an interview with Robert Sabin, "He remembered a song, 'Follow the Footsteps in the Snow' which accompanied a sort of snowflake ballet danced by a group of women. In addition, at the Marlborough Theatre, he saw Loie Fuller or one of her imitators in a spectacular 'Fire Dance,' 'Golden Dance,' and a vision called 'Under the Sea.' He enjoyed tableaux vivants and 'Grecian Plastique,' and later on, as a young man, he stood four hours in the cold to buy a gallery seat for Anna Pavlova."[3] Seeing Pavlova in *The Fairy Doll*, in 1924, was an important moment in his life that contributed to his desire to become a dancer and choreographer. His interests in listening to and playing music, reading literature and philosophy, going to art galleries, and especially acting occupied whatever free time he had in his later teenage years.

Drawn to theater, he joined several acting societies, where he enjoyed the challenge of learning various kinds of dramatic character roles. He first participated in the acting society attached to his local church and, subsequently, the City Literary Institute, similar to Cooper Union in New York City. Finally he worked with St. Pancras People's Theatre, which staged a different play every weekend. As a result of this acting experience, he developed a passion for Shakespeare. It was in this same district of St. Pancras that George Bernard Shaw acquired the practical experience as a vestryman that was to serve him so richly in his plays. After a second year at St. Pancras, Tudor was invited to make the dances for the Christmas show, which, as he remembered them, consisted largely of "swinging a girl around violently to Grieg's 'Hall of the Mountain King.'"[4] In order to prepare himself for choreography, Tudor read voraciously, especially the dance magazines that in those days contained many "do-it-yourself articles," with instructions for learning the tango as well as the *entrechat*.

Although Tudor had bookish tendencies, he did not attend university. When asked why, he responded disconsolately that he had failed the entrance examination, and that he felt very sad at not being able to continue his education. Unsure of what he wanted to do with his future, one year after graduating from high school he took a theology class in an old, cathedral-like building on the Strand. He studied New Testament Greek and was thinking seriously about the ministry. One of his tasks was to stand at a lectern in front of the class and give a brief lecture. It was not to his liking, and he realized then that religion would not be his profession. For a while, he also contemplated becoming a musical conductor, "a very powerful sort of position. . . . And they can swing the emotions and have power over all those little people around them."[5]

Tudor was already twenty when he found his way to Marie Rambert and ballet. Having taken Spanish dancing as well as other kinds of dancing classes, he believed that his appearance as an actor would be enhanced by the studied formalism of ballet. He hunted out the Charing Cross bookshop in London where Cyril Beaumont, the writer and critic, worked, and asked him for the name of a good ballet teacher. Beaumont recommended Margaret Craske (with whom he later developed a lifelong working relationship) and Marie Rambert, who first nurtured his gift for choreography.

Marie Rambert, or "Mim," was born Miriam Rambam in Warsaw in 1888 to a Jewish family, who changed their name to Ramberg as a step into a wider secular world. She grew up in an environment where reading and literature were highly valued. While she was a schoolgirl her inclinations grew increasingly toward personal and political freedom in an atmosphere seething with revolutionary ideas. At age seventeen, she was sent to Paris to perfect her French and eventually to study medicine, but instead she came under the influence of several writers, actresses, and dancers, including Colette and Raymond Duncan, Isadora Duncan's brother. She saw Sarah Bernhardt's last performance, was thoroughly astounded by Isadora's concerts, and in 1906 traveled to Vienna to see Ruth St. Denis' experimental work.

Rambert threw herself into the rich creative life of prewar Europe. Gradually, she developed her own interest in making improvisational and *"intime"* dances and decided that summer classes in rhythmic music would help her become a better choreographer. She planned to attend Emile Jaques-Dalcroze's music courses for only a short time, but stayed with him for three years, becoming fluent in his eurhythmic principles, a system of musical training based on rhythmic movement.[6]

In 1913, when Rambert was still working with Dalcroze, the impresario Serge Diaghilev and his choreographer Vaslav Nijinsky, visited the school, which had moved to Hellerau outside Dresden. Dalcroze recommended that Rambert teach Nijinsky basic principles in rhythmic training to help him decipher Stravinsky's complex score for *Le Sacre du Printemps.* (She later discovered that Dalcroze's own interpretation of musical scores could be rigid and limiting.) Eventually, Diaghilev took her into the corps de ballet of his Ballets Russes company, where she acquired invaluable experience as a dancer and an artist.

It is important to remember that, from 1911 on, Diaghilev's ballet made regular visits to London, bringing the excitement and exoticism of a transplanted Russia to the bleak London ballet scene. Until the death of Diaghilev in 1929, the Ballets Russes steadily spread its radical aesthetic and contributed to changes that ultimately prepared the way for Rambert's ballet company. They cultivated a mystique of fashion, fascination among chic literati, and a following by aristocratic dandies. Lynn Garafola has traced London audience peripatetics in the last chapter of her book *Diaghilev's Ballets Russes.* From 1918 to 1922, intellectuals adopted the Ballets Russes as their own: "What delighted Bloomsbury above all, was the visual modernism of the troupe's newest productions. . . . Diaghilev used his proscenium to frame genuine works of art; he turned the stage into a magnificent gallery serving the artists Bloomsbury had championed for a decade."[7] Avid aristocrats, an all-male university élite, and eccentric literati joined in their support. For example, a gossip column spoke of one of the premieres: "A special and very sophisticated audience always haunts the Russian Ballet, and on the first night at the Princes the 'regulars' turned up in full force. The Sitwell brothers were accompanied by Mrs. Sacheverell Sitwell who looked elegant in a gold brocade coat."[8]

In addition to these select groups, Diaghilev attracted more of a popular crowd than the one he excited in Paris. Though he did not actively solicit spectators of all classes, "he won a British following that swept across the social spectrum. By promoting modernism through the music halls, he added to the old élite an audience of impassioned intellectuals and modest pleasure-seekers."[9] Diaghilev also arranged performances at large variety theaters, such as the Coliseum, Alhambra, and Empire, rather than the Drury Lane and Covent Garden. As a result, more varied groups of people filled the seats.

By the time Rambert began to resuscitate English ballet, a place in the hearts of Londoners was ready for her. She settled in England in 1914 and gradually realized her dream, which was to open a school and run a company with the same vitality and innovative qualities as the Ballets Russes. In 1926 she was able to form a small company with dancers and choreographers she had trained. The troupe evolved into the Ballet Club in 1930, and in 1935 it

Members of Rambert's Ballet Club ca. 1931. Standing, left to right: Prudence Hyman, Frederick Ashton, Antony Tudor, Alicia Markova, Andrée Howard, Marie Rambert, Robert Stuart, Pearl Argyle, Diana Gould, Maude Lloyd, Elisabeth Schooling, and Betty Cuff. In front, left to right: William Chappell, Rupert Doone, and Suzette Morfield. (Photo courtesy of the Rambert Dance Company Archive)

became Ballet Rambert. The Ballet Club, as described by Agnes de Mille, who was one of its early participants, "functioned in what had been the vestry house of a small odd-shaped church situated in an apex of land on the Notting Hill Gate bus routes." Inside it was "a geometric conglomeration of boxes, hallways, levels, closets, little auditoriums and stairways"; and the centerpiece, the theater itself, "had the air of a tiny eighteenth-century princeling's court theatre."[10] Here the Ballet Club gave performances on Sundays, while weekday evenings were reserved for plays produced by Rambert's husband, the playwright Ashley Dukes, and the talented group of actors, playwrights, and set designers he had gathered around him. The company flourished and was considered important enough to participate in a new venture in English dance, the Camargo Society.

With the demise of Diaghilev's Ballets Russes in 1929, something was needed to replace its energizing London seasons. The Camargo Society was conceived in a Soho restaurant called Taglioni's by the young Arnold Haskell and P. J. S. Richardson, the editor of *The Dancing Times*. The declared aims of the Society were: "To produce original and classic ballets before a subscription audience at a West End theater four times a year on Sunday evenings and Monday afternoons. And to invite for these productions the collaboration of eminent composers, painters, and choreographers. As well as to engage for the performances the best dancing talent in London and abroad."[11] The Society, small but well-supported, presented its first production in the newly built Cambridge Theatre on October 19, 1930, and its last in 1933. During its three-year existence, it admirably provided support for Rambert and her crop of dancing, choreographing protégés.

It was this milieu that Tudor stepped into. After speaking with her on the phone, Tudor fearfully went to see "the great Rambert" on a cold winter's day near the end of the year in 1928.[12] He told her that he could not come to classes before 4:00 p.m. because he was working from 6:30 in the morning to 3:00 at the Smithfield Market. Rambert was "impressed by the fact that he was prepared to study dancing immediately on top of a long working day."[13] In order to earn his keep with Rambert, he played piano and taught the elementary classes, was her stage manager, electrician, and lighting person as well as a performer, and acted as secretary of the Ballet Club, for which he was paid two pounds a week for keeping very good accounts. Rambert also offered him a place to sleep. "When I left home, my parents probably wondered why on earth I hadn't left earlier, although it was never mentioned. I moved all the way across town where there was a little room off Rambert's studio where I could live."[14]

Tudor enjoyed studying at Rambert's in part because the studio was located in a posh area of London: "I was with all sorts of snobbish students, and I rather liked that."[15] A later commentator has reported that those early days were not easy:

> The school was a strange world to him at first . . . their snobbishness was aroused by his accent and his obvious unfamiliarity with the fashionable world. He was excluded from their parties. But he was far too busy to waste time resenting such slights.[16]

He was listening with care to Rambert's suggestions about what books he should read, how he should change his name (which he did in 1930), how he should take private ballet lessons with Pearl Argyle and Harold Turner in order to catch up with the other male dancers in the Ballet Club company, and how he should eradicate his unacceptable cockney accent. This last was accomplished "on Wimpole Street (where all the surgeons lived) with a fellow who cost a fortune for a half hour to an hour. All he did was to make me open the London *Times* to the editorial column, put on music and simply said, 'read it.' And reading the editorial column of the London *Times* is absolutely horrible, but after six sessions the accent was gone."[17]

Tudor's fortuitous meeting with Marie Rambert began one of the more important mentorships, however brief, in ballet history. She offered him financial security and a window on the world of theater and dance that totally captivated his poetic spirit. No doubt Rambert saw possibilities in this working-class boy; she certainly had much to do with William Cook's transformation into "Antony Tudor."

During his early years with Rambert, Tudor was often invited to dance with various ballet companies, including Ninette de Valois's Vic-Wells ballet company, in spite of his inexperience. Male dancers were at a premium; Tudor noted that there were "probably fewer than a dozen males in the entire United Kingdom who might, in their vanity, dream of having a working future in the insecure, alien world of the ballet."[18] He started performing immediately, appearing in 1929 as "slain lover" to choreography by Penelope Spencer in an English opera, *Cupid and Death,* at the Scala Theatre in London. In 1931 alone, he danced for the Camargo Society in Frederick Ashton's *Façade, The Lady of Shalott,* and *A Day in a Southern Port,* and even partnered Rambert dancer Prudence Hymen in one of the entries in "Aurora's Wedding." (According to Maude Lloyd, the production was underwritten by John Maynard Keynes, the noted economist and husband of the esteemed Russian ballerina Lydia Lopokova. They brought together the Vic-Wells and Rambert companies with Tamara Karsavina.)

Even more beneficial to Tudor than his grooming as a dancer, was Rambert's indefatigable discovery and promotion of young choreographers. One of those was Frederick Ashton. He had begun studying with Rambert in 1924 and made his first ballet for her in 1926. Although he also studied with Serafina Astafieva, Nicholas Legat, and Margaret Craske and later at Ninette de Valois's Academy of Choreographic Art, he continued to be Rambert's pupil because she offered him free lessons. Furthermore, through her former connection with Diaghilev, she was able to take her students *gratis* to Ballets Russes performances. There, the new ballets of George Balanchine, Léonide Massine, and Bronislava Nijinska had a great influence on Ashton's early work. He became a regular choreographer for the Ballet Club until 1934, when he decided to join de Valois's Vic-Wells Ballet.

Tudor was introduced to Ashton's work when Rambert took him to a Ballet Club performance. Whether or not there was an immediate impact, when Tudor began to choreograph, he and Ashton became rivals of a sort. Ashton had a lively imagination and was particularly adept at creating ballets with profoundly lyrical as well as ironic qualities. Not only were traits shared, but themes that Tudor would eventually take up for his ballets were also successfully used first by Ashton, such as in his *Passionate Pavane* and *Mars and Venus.* Throughout his years working in England, Tudor always seemed to be several steps behind Ashton.

In the beginning, however, Tudor had everything to gain from his association with the Ballet Club and with Rambert. She introduced him to the talented personalities who surrounded her husband at his Mercury Theatre, where Tudor met people such as Nadia Benois, who came to design the stage set for his *Descent of Hebe* and *Dark Elegies;* and Hugh Stevenson, who designed *Jardin aux Lilas* and *The Planets.* Tudor recognized his advantage: "To be accorded even a peripheral acquaintance in the artistic circle with which her husband, Ashley Dukes, and she surrounded herself was an honor and useful."[19]

But the situation was not always blissful. Rambert had a volatile temper and an overpowering personality, which her protégés eventually found to be suffocating. She taught, Tudor said, "exuding energy and command, and alternating between yelling and whistling and laughing."[20] It was understood that as soon as most of her students could get away, they did. For Tudor, it was a love-hate relationship.

As a counter to this intemperance, Tudor found a powerful ally in a newcomer to the Rambert dancers, Hugh Laing Skinner. In 1931, Laing had traveled to London from his home in Barbados to study painting at the Grosvenor School of Modern Art. He soon gave that up to become a ballet dancer, which he accomplished by studying with Margaret Craske and Marie Rambert. His first performance with the Ballet Club was in 1932, a year before his appearance in Tudor's ballet *Atalanta of the East.* By then he and Tudor were already good friends and colleagues. His gift for costume design was to be crucially important to Tudor. (He designed the dresses for *Judgment of Paris* and assisted in the designing of *Gala Performance, Pillar of Fire, Shadow of the Wind* and *The Divine Horsemen*—but union rulings prevented his receiving credit.)

Laing, a fierce supporter of Tudor's, would often get into screaming sessions with Rambert, usually defending one or another of Tudor's unorthodox ideas. Laing need not have fought with such ardor, as, much later, Rambert mellowed with the passage of time—when asked in 1966 about Tudor's contribution to English ballet, she quickly responded, "Tudor did things which no one else had attempted at that time. . . . He expressed the most delicate human relationships in a very simple gesture."[21]

In 1931, Tudor's aspiration to be a serious choreographer came to fruition. He began to work on a ballet that would explore Italian art and Shakespearean themes. Fernau Hall, the late dance critic and Tudor champion, stated with excess that Antony Tudor, "the Jules Perrot of the 20th-century ballet," created his first ballet only six months after he began seriously studying the technique with Margaret Craske and Marie Rambert.[22] In fact, Tudor did not start studying with Craske for a while and, according to Rambert, "he was . . . in his third year when he chose the amusing letter scene from *Twelfth Night* to be the central focus of this first ballet."[23]

Cross-Garter'd (November 12, 1931)

Cross-Garter'd was presented on the tiny Mercury Theatre stage during the third season of the Ballet Club (November 12 to December 2), on a program with Ashton's *Mercury, The Lady of Shalott,* and *Mars and Venus,* and Susan Salaman's *Le Rugby,* with Petipas's *Lac des Cygnes.*

To prepare for his ballet, Tudor traveled to Italy to visit the museums in Florence, using money his father had given him as a birthday present to buy a bicycle. Tudor felicitously chose the music of Girolamo Frescobaldi, who, it seemed to Rambert, was a most appropriate choice since he was an Italian contemporary of Shakespeare's. In deciding on the seventeenth-century artist Ludovico Burnacini as the inspiration for his costumes and sets, Tudor was choosing an exaggerated, Plautine comic design. Riotous in a Rabelaisian fashion, Burnacini represented the spirit of the baroque grotesque.

The scenario of *Cross-Garter'd,* drawn from Act II, scene v, and Act III, scene iv, of *Twelfth Night* concerns Malvolio, the pompous, conceited steward to the Countess Olivia. His wildest desire would be to marry Olivia. Maria, Olivia's shrewd serving woman, has designs on Sir Toby Belch,

Cross-Garter'd, *1931. Left to right: Elisabeth Schooling, Maude Lloyd, Betty Cuff, Rollo Gamble, Prudence Hyman, Walter Gore, Antony Tudor, and William Chappell.* (From a photo by Pollard Crowther, courtesy of the Rambert Dance Company Archive)

Olivia's fat, jolly, hard-drinking cousin. Toby's companion, Sir Andrew Aguecheek, is a wealthy, skinny, rather feebleminded knight who has come to woo Olivia. And Fabian symbolizes the typical sly, merrymaking Elizabethan servant, the sort of character whom higher-born aristocrats like Toby and Andrew enjoy spending time with.

The scene transports us to Olivia's garden where Maria, Sir Toby, Sir Andrew, and Fabian are preparing to conceal themselves in order to observe Malvolio's reaction to the trick Maria is going to play on him. They gleefully hide behind a "box tree," laughing and excitedly predicting the success of her scheme. She tosses a letter down on the garden path where Malvolio, who has been strolling and preening himself nearby, is sure to find it.

Malvolio enters, deep in thought, in the middle of a hilarious soliloquy punctuated by bursts of outraged asides from Toby, Andrew, and Fabian. Malvolio daydreams about his innate suitability as a husband for Olivia; about how, after being married to Olivia, he would call for "his" cousin Toby in order to chastise him for his wild behavior and his foolish friends. He suddenly picks up the letter, identifies the writing as "my lady's hand," and reads the beginning, which starts with a rhymed riddle, "Jove know I love. / But who? Lips do not move. / No man must know," and continues, "Some are born great, some achieve greatness, and some have greatness thrust upon them. . . ." The letter, with its wonderfully funny instructions to its reader and its clever parody of Malvolio's own style, is a testament to Maria's superior wit and cunning.

Sometime later, following precisely the instructions of his mysterious correspondent, Malvolio appears wearing bright yellow stockings with garters wrapped around them in a zigzag pattern (cross-garter'd). He greets Olivia with a cheery, "Sweet lady, ho, ho." There ensues a long, comical exchange between the two, during which Malvolio misinterprets every one of Olivia's remarks in terms of the letter he has supposedly received from her. When she tells him that she sent for him because she was feeling sad, he replies that he supposes he could be sad, too (since his garters are so uncomfortably tight), but continues with a meaningful leer that "if it please the eye of one" it is all right with him. When she asks if he wants to go to bed (because he is sick) he gleefully misunderstands her solicitude as an invitation to him. Olivia is appalled by his behavior, and she is certainly not enlightened when he answers with knowing winks and references to the letter. With a few distracted words to Maria—"Let this fellow be looked to. . . . Let some of my people have a special care of him," she rushes off, commending Malvolio to the special care of her cousin Toby.

Malvolio has so allowed his egotism and his ambition to run away with him that by now he is completely out of touch with reality. He does not even notice that Olivia's reaction to his lascivious advances is compassionate, rather than passionate. Nor can he imagine the unlikelihood of her ever addressing such a love message to him.

Rambert was concerned about Tudor's choreographic abilities until he showed her the sketches for a solo and a *pas de deux*. It took him one week—working from a series of precisely written notes in which he mapped out the patterns, the gestures, and, according to Maude Lloyd, "not quite a step a beat"[24]—to devise Malvolio's entrance, "duly pompous," and a silly *pas de deux* for Malvolio in his outrageous costume and an amazed Olivia, danced by the "beautiful Maude Lloyd." She wore a black velvet costume that provided the proper weight, luxury and charm; she did not wear pointe shoes.

The critic Tarquin Hobbes wrote:

> Tudor's first sketch of a dance for himself as Malvolio seemed somewhat lacking in character and had no special quality to distinguish it from the movements of Fabian, for example. But this dance of Malvolio's was to be done with his traditional tall stick, and Tudor thought that this would completely change the character of the movements; and he was right.[25]

Maude Lloyd recalled that Tudor created most of the movements in *Cross-Garter'd* from the ballet technique. She particularly remembered Tudor's invention of a new step where she continued a *rond de jambe à terre* with one leg while pivoting on the standing leg. He used tiny little Renaissance steps for Maria, danced by Prudence Hyman, and for the men, running, skittering steps on half-pointe, while the gesturing was derived from paintings of commedia poses and tableaux.[26]

With *Cross-Garter'd,* the public immediately sensed the launching of an important career. Even Léonide Massine, who rarely said anything about anyone, complimented Tudor backstage after seeing *Cross-Garter'd.*[27] On the other hand, Marie Rambert noted cursorily in her autobiography, "It was not a good ballet but it demonstrated enough talent to merit its inclusion in the season of 1931."[28] Likewise, Elisabeth Schooling, who had perfomed in it, recalled succinctly that it was "much too busy."[29]

In several reviews, Tudor's merits as a neophyte choreographer received praise, especially for the clarity and charm of his vision of the garden scene of *Twelfth Night*. Lionel Bradley predicted that Tudor would contribute important and creative ideas to the London dance scene. "But with *Cross-Garter'd,* Tudor showed from the outset great originality of conception and novelty of choreographic language which developed steadily through the many works he composed. . . ."[30]

The Dancing Times added:

Mr Tudor has an excellent idea of design in detail, but the ballet regarded as a whole lacked a definite shape. The dances gave the impression of having been most carefully and painstakingly arranged, and if this really be a first attempt Mr Tudor, as a choreographer, should be given every encouragement. Incidentally he himself made a striking Malvolio and he received great help from the other members of the company.[31]

A contemporary newspaper displayed a picture of the four women rehearsing *Cross-Garter'd* for a midnight ballet party to aid Queen Charlotte's Hospital. Thus the production proved itself entertaining, a good vehicle for showing the company in its best light and appealing to the upper crust whom the savvy Rambert was trying to cultivate.

Cyril Beaumont noted, *"Cross-Garter'd* was of particular interest as the first attempt of a new aspirant to choreographic honours."[32] Beaumont praised Tudor for "working out his ideas in his own way," using the letter scene in Olivia's garden as his frame. "The steps were simple, the mime simple but it was all done in the lusty spirit of the text."[33] Beaumont found the opening scene appropriate but too long, and said that in the last scene the dancers were too impetuous as they lifted up their legs in a way "that would never have been permitted in the dance of Shakespeare's day."[34] It must be said that Tudor heeded this criticism, as he not only became very interested in Renaissance and baroque dances but was careful to restrain leg extensions in any of his historical endeavors.

Fernau Hall reflected, rather prejudicially, that when Tudor put on his *Cross-Garter'd*, "the world of ballet and his world of the Ballet Club were entirely under the influence of the Russian ballet and Diaghilev with their clever, clever, and frivolous *'étonnes-moi'* credos. Tudor's aesthetic did not match this point of view."[35]

Lionel Bradley lauded Tudor's satirical spirit in his remarks about a later production:

> *Cross-Garter'd* is a clever and witty depiction of the letter incident from *Twelfth Night*. . . . with its presentation of grief and dignity, followed by shocked bewilderment at the tranformation of Malvolio is nicely depicted with just sufficient of a satirical edge to it. Gore's Malvolio is another good portrait and the rest make an able background with Frank Staff's fantastical "Sir Andrew" and Sally Gilmour's pert "Maria" as the highlights.[36]

Cross-Garter'd was in the repertoire of the Rambert Ballet Company and played in good health until 1938.[37]

Three months after the premiere of *Cross-Garter'd*, the neophyte choreographer Antony Tudor presented two more ballets, *Mr. Roll's Military Quadrilles* and *Constanza's Lament*.

Mr. Roll's Military Quadrilles (February 1932)

The *Quadrilles,* which was performed several times at the Ballet Club, appeared on different bills with Ashton's *La Péri, A Florentine Picture, Divertissement, Saudade do Bresil, Pompette; The Tartans* or *Dances on a Scottish Theme;* Tudor's *Constanza's Lament;* Susan Salaman's *Le Rugby;* and Fokine's *Les Sylphides.*

At a conference on Antony Tudor for the Sadler's Wells Theatre Community and Education Project in 1987, John Percival recalled that *Mr. Roll's Quadrilles* was the second piece Tudor created, probably because there were some old costumes around that seemed appropriate. "Tudor used them to make a little ballet which was simply dances. So there, right at the beginning of his career, he went from narrative ballet (*Cross-Garter'd*) to plotless ballet in one stride, and it wasn't very long before he did something more interesting."[38]

Maude Lloyd exclaimed in a recent interview, "Oh, *Mr. Roll's Quadrilles* was a parody. Mim didn't like it very much." Lloyd described the quadrilles as "knockabout classical" to the kind of music one might hear in music halls. "Mr. Roll's Quadrilles (the music, not the ballet) is an old English piece of music which [Tudor] thought amusing. It was for Prudence Hyman and Tudor and me."[39]

Constanza's Lament (February 4 or 11, 1932)

The ballet played on the same program with Balanchine's "The Shepherd's Wooing" from *The Gods Go A-Begging* with music by Handel in a duet for Pearl Argyle and William Chappell, as well as with the ballet *Pompette,* arranged by Frederick Ashton and danced by Andrée Howard. The original French title for this choreographic comedy in one act was *Les Femmes de bonnes humeur,* with a libretto and choreography by Léonide Massine and set design by Leon Bakst. It was premiered by the Ballets Russes on April 12, 1917, in Rome. Based on Carlo Goldoni's play, it tells of a complicated love affair between Constanza and Rinaldo, who scheme and overcome many obstacles resulting from their parents' opposition to their marriage.

From this story, Tudor created a simple, though very interesting, solo for Diana Gould called *Constanza's Lament.* Tudor put together a simple pointe piece, in which Constanza is dressed in a wig and an eighteenth-century ballroom dress that is not quite authentic, as the hem is way above the ankles. The miserable Constanza floats from one part of the stage to the other, seemingly wafted here and there by the body's sobbing thrusts. The steps that created the impression of Constanza's drifting and longing consisted largely of *bourrées* with *renversés* (dips of the body front and the lifting of the leg to the side with the body turning around itself) and *balançoires* (swinging the leg front to back in arabesque). Opposition and movements on the diagonal figure importantly, while the arms and hands move softly through movements of the wrists. During the *bourrées* across the stage from one side to the other, the hands and upper body form shapes that dissolve successively for emphasis and accent. In developing a short solo for Constanza that needs no explanation, Tudor experimented with characterization and tone.

Tudor was rehearsing and dancing a number of roles at this time. He appeared as the Trainer in Susan Salaman's *Le Boxing,* one of her *Sporting Sketches* in February 1932; in Frederick Ashton's *The Tartans, Dances on a Scotch Theme;* and in Rupert Doone's *The Enchanted Grove* as a Warrior on March 11, 1932, at the Sadler's Wells. The March 1932 *Dancing Times* mentions that Antony Tudor was also dancing in a musical, "Butterfly Ballet" with Gina Malo.

The situation of ballet in London at the time revolved around Rambert's Ballet Club and Ninette de Valois's Vic-Wells Ballet. De Valois also fought hard to develop the excitement and audience-pleasing works that had characterized Diaghilev's Ballets Russes, which she, too, had danced with. Both she and Rambert as directors struggled to create their personal visions and were determinedly looking toward the future. Both of them encouraged choreographers, but when Ashton returned to de Valois, he became the focus for new choreography.

Tudor would have given a great deal to choreograph for de Valois, but, as David Vaughan has noted,

> she was dubious; in the first place, she felt that Tudor needed the kind of experience of working with great choreographers that she had had with Diaghilev and Ashton with Ida Rubenstein,[40] and advised him that there would still be an opening for him at the Wells at the end of that time, and the opportunity to choreograph, but Tudor rejected her advice and in the next two years made two mature, important ballets for Rambert, *Jardin aux Lilas* and *Dark Elegies.* Also, perhaps more importantly, she realised that Tudor was not the classic choreographer she needed.[41]

Rambert had a more democratic, serendipitous attitude than de Valois, encouraging anyone who truly desired to excel in making dances. According to Kathrine Sorley Walker, what was unique about Rambert's company was that "the idea of turning choreographer was always talked about. Ashton had been the first one launched by Rambert, and Susan Salaman was the second, with a children's ballet in 1929. Susan's talent for composition was considerable. In all, she contributed eight pieces to the Rambert repertoire, showing an ability to produce fresh and witty dances. Her career was regrettably ended by illness after *Circus Wings* in 1935."[42]

Tudor was the next to come along. No doubt encouraged by the welcome from the critics, his next ballet appeared a mere six weeks after the premiere of *Mr. Roll's Military Quadrilles* and *Constanza's Lament.*

Lysistrata (March 20, 1932)

Lysistrata was suggested to Tudor by Ashley Dukes, who also devised the subtitle, *Strike of Wives.* It opened on a program with *Pompette,* a *pas seul* choreographed for Andrée Howard, and *Aurora's Wedding, Mars and Venus,* and *Façade,* all by Frederick Ashton. (Although Tudor appeared in many of Ashton's early works, Ashton never danced in Tudor's ballets, as he was obliged to perform with Ninette de Valois's Vic-Wells company.) Tudor costumed the company in pseudo-classic finery, and Tudor matched the costumes and scenery with some pseudo-classic movement that he saw on the Greek vases and friezes in the British Museum.

Based on the original comedy written in 411 B.C. by Aristophanes, *Lysistrata* presents the Athenian women refusing to perform their wifely duties until their husbands forswear war. Before proclaiming her plans, Lysistrata has the older women seize the Acropolis in Athens in order to control the treasury. The Spartan men, unable to endure prolonged celibacy, are the first to petition for peace, on any terms. Then, Lysistrata, in order to hasten the war's end, has a nude girl exposed to the two armies. Thereupon the Athenians and Spartans, goaded by frustration, make peace quickly and depart for home with their wives. With only minuscule program notes and an incomplete series of photographs surviving, it is difficult to know the complete scenario of Tudor's version of *Lysistrata.* Critics however, emphasized that Tudor's wonderful characterizations and pantomime suited the play's important moments. As in *Cross-Garter'd,* Tudor was comfortable with a theater script in which the personalities were already well developed and described in words.

He was developing rapidly as a choreographer, and every single role in *Lysistrata* was sharply characterized by his wonderfully intelligent dancers. It survived eight years in the repertoire.

Rambert mentions in her autobiography that one of the best moments in the ballet occurred when Lysistrata first calls upon the wives to strike against their husbands. There followed a *pas de deux* with Alicia Markova

Lysistrata, *1932. Maude Lloyd as Lysistrata and the Group.* (Courtesy of the Rambert Dance Company Archive)

and Walter Gore, in which the spectacle of Gore minding the baby, who would not stop yowling, was most hilarious. Markova, who played the wife Myrrhina, recalled that, unlike Lysistrata, Myrrhina is tormented by her sympathetic feelings for both her husband (Walter Gore) and her baby (a doll). She decides, however, to take vengeance against her warring husband by using the baby in a calculated plot. During her *pas de deux,* she rushes from her husband to the baby, whom she had placed on the floor near her husband. But the baby keeps wailing, thwarting any intimacy between her and her husband. Markova appreciated the fact that her role demanded an approach different from one she was accustomed to using in other works that seemed to have more surface frivolity. She said that Tudor brought her "down to earth."[43]

The choreography had real comedy, but Rambert did not appreciate Tudor's use of the Prokofiev score, which seemed unsuitable to the story. She noted that Prokofiev himself came to the ballet and "did not hide his disappointment."[44] In 1938, Tudor chose another score by Prokofiev far more successfully when he presented *Gala Performance* to Prokofiev's Classical Symphony.

35

Hugh Laing commented that Tudor began to feel a new way of working after choreographing the solo for Walter Gore with the baby:

> It was full of humor and full of humanity. He'd put it down and walk away and then it cried. And the wife watches from a distance and laughs at him. And this was a human relationship. Because it was such a human little solo which was beautifully arranged. And I think it may have given him the cue to acting . . . the responsibility to humans and not to cardboard figures.[45]

After the men are emphatically told by their wives that they are on their own, Tudor rewards them with either a kitchen tool or a baby and they dance a trio describing their woeful state. The picture of a warrior nursing a baby or performing household chores remains a moral statement for all time.

Maude Lloyd, who took over the role of Lysistrata from Diana Gould, made a vigorous leader of the feminist revolt. She recognized that Tudor always created complex characters, even his comic ones, so that no one appeared flat or superficial. Lloyd felt that her Lysistrata benefited from expressing the dual qualities of strength and female wiliness. She remarked that Lysistrata was not particularly easy for her, as she was essentially a lyrical dancer, whereas Gould was tall and rangy, more at home with an aggressive role. Lloyd recalled that during a powerful duet with Tudor (playing her husband), a real fight ensued in which she actually kicked him; she mentioned that the scene exhausted her. Here again, Tudor's interest in the theater, in the reality of personification and gesture, help communicate the irony as well as the obvious comic features of this ancient play.

Laing added:

> It was made for a very small stage, the Mercury, eighteen feet wide . . . the Prokoviev music was delightful. The husband who comes back and his wife says, "All right. You look after the baby." She would have nothing to do with him. And he does this fascinating dance trying to put the baby to sleep. It was sort of a stuffed little nothing, you know, and he was utterly wonderful, just wonderful.[46]

Lionel Bradley's *Journals* contain a detailed description of the February 9, 1937, performance of *Lysistrata*, five years after its premiere. He described the colors of the costumes: "Lampito's costume was a sort of French Revolution red and she wore a black cross on her cheek." Bradley emphasized the moment when the husbands return and are astonished to be met with kicks and blows instead of kisses. Elisabeth Schooling wrote that "Tudor wanted a bed represented by a sheet on the stage. Rambert would not have it in her theatre—ensuing row."[47] *Lysistrata* survived eight years in the repertoire.

During the same month Tudor was rehearsing other jobs and shows.

He had choreographed *In a Monastery Garden* in March 1932, a film shot in the Twickenham Studios, with other Rambert dancers. In October, Tudor danced a passport inspector in de Valois's *Douanes,* and six days later danced the role of Edwin Morris in Ashton's *The Lord of Burleigh* at the Old Vic Theatre. He also took the role of Apollo in Ninette de Valois's *The Origin of Design.*

In the late fall of 1932, Tudor was living in Chiswick with Hugh Laing in a mews off Campden Hill. When Agnes de Mille, American dancer and choreographer, was first introduced to them by Rambert, she saw "two young men in sailor pants bent over and painting scenery. They straightened up and regarded [me] in the fading light."[48] De Mille had been invited by Dukes and Rambert to do a series of solo recitals at the Mercury and at the same time was offered the opportunity to take a class with Rambert at her own expense. Instead, she studied with Tudor, loving the unusual, more personal way that he taught. In the six years that de Mille performed solo concerts, Tudor stage-managed every one of them, receiving two pounds per show, while Hugh Laing assisted her as partner in these ventures.

A third attempt by Tudor to create a story ballet seems to have brought him some of the recognition he aspired to.

Adam and Eve (December 4, 1932)

The evening was produced by the Camargo Society, with *Les Sylphides* and Penelope Spencer's *The Infanta's Birthday* also on the program.

As in *Cross-Garter'd* and *Lysistrata,* Tudor's talents were demonstrated in the way he portrayed his characters and in the consciously heightened and amusing manner of his treatment of small groups of dancers onstage. Tudor's developing ideas of choreography were based on ballet technique, but with a peppering of dramatic accents and gestures. The scenario approximately follows the biblical story, with the important difference that a final reconciliation scene of all the characters in the ballet was added.

Constant Lambert's music was originally composed in 1925 as a *suite dansée,* in which form it had been submitted to Diaghilev. Diaghilev declined the scenario, possibly because at that moment he needed a ballet with a larger cast, but he earmarked some of the music for inclusion in *Romeo and Juliet.* For *Adam and Eve* it was restored to its original form and sequence. It is in three scenes, each comprising three movements, with the scenes being connected by the same interlude, as follows: Scene 1—Sinfonia, Siciliana, Sonatina Intermezzo Pastorale; Scene 2—Burlesca, Musette, Tocatta, Intermezzo Pastorale; Scene 3—Rondo, Sarabande, and Finale.

Burlesquing in the wittiest way, Tudor places this unauthorized version of the Adam and Eve tale in a futuristic Garden of Eden. "Tudor approached the ballet lightly but not flippantly, using the flavor of the commedia dell' arte."[49] With rather surrealist imagery, the story mocks the tale of the apple and the birth of original sin instigated by a serpent with Don Giovanni airs:

> The stars, Adam and Eve, looking like sun-bathers, are introduced to each other (Adam looks like a footballer and Eve like a soubrette) by the Fowls of the Air, with the approval of two Seraphim, there entered the Serpent (A. Tudor). When Grigorova was half way through her angel's duty of driving the erring pair from the garden, she relented. The serpent had another idea. He seized the avenging angel with her flaming sword and forced her to share a bite of the apple with him. All ended happily with a double marriage conducted by the seraphim. Tudor's wry sense of humor and his "jeu d'esprit" carry the story of its fruitful ending, as even the apple is given an effective part.[50]

What role could be more Tudoresque than the serpent in the Garden of Eden? And who with the glint in his eye and his diabolical tongue could best play the serpent? "This was the greatest success of the evening. Tudor appeared as the Serpent, a masked Harlequin in black and white."[51]

Several of the critics acknowledged that, though the tone and the scenarios were entirely different, Anton Dolin's performance of Adam reminded them of *Job,* a ballet by Ninette de Valois in which Dolin also wore a loincloth, but in *Adam and Eve* Tudor appropriately added a large figleaf. Arnold Haskell praised Dolin's performance, "at ease, witty, and superb in control. No more magnificent impersonation of our grand ancestor than Dolin's has ever walked, run or leapt. Prudence Hyman was worthy of her Adam and her origin. She is the most gifted of all our young dancers, fresh, fascinating with a strong personality."[52]

But *Time and Tide* disagreed with Tudor's interpretation and skill:

> He "improved" on the original in a final tableau in which the Angel with the flaming sword is induced to taste of the fruit of the Tree of Knowledge. . . . The choreographer has failed to realise his comment in the terms of his medium and though wit is implicit in grouping and situation, it is absent from the pas. Not even the art of Dolin could succeed in infusing the role of Adam with significance.[53]

A newspaper feature of the day emphasized the English feeling of artistic competition with the French:

> The whole evening was wickedly irresistible. Now that the Camargo Society has shown what British ballet can do when it tries, the next thing should be to

send the latest sample to Paris and show the French people what England is up to in the twentieth century. It would surprise them.[54]

Arnold Haskell in *The New English Weekly* expressed his reservations about the ballet, centered on John Banting's "infantile decor, a crude imitation of Max Ernst at his worst." Haskell then alluded to a remark that Dolin's mother had made after the performance "that she had never before seen such a lascivious and suggestive ballet. Tudor as the Serpent, dressed as Harlequin was a real creation, not the conventional squirms in the manner of Uriah Heep, comedy and criticism at the same time. Tudor took risks; his version succeeded."[55]

On February 7, 1933, Tudor played a "rural dancer" in Ninette de Valois's *The Birthday of Oberon* at the Sadler's Wells Theatre, and in May, he presented his next ballet. It was the most ambitious so far.

Atalanta of the East (May 7, 1933)

Appearing on the same program with *Atalanta of the East* were Fokine's *Les Sylphides* and *Pavane pour une infante défunte,* and *Les Masques,* with choreography by Frederick Ashton.

The program reads: "Ballet in two Scenes—1) Vikram prays to the Goddess to help him in his quest of the fleet footed Sita and receives from her the golden apples. 2) Sita, secure of victory, accepts the challenge of Vikram, but is vanquished by the ruse of the golden apples which Vikram drops in the race."

Tudor's lifelong interest in Eastern themes for his ballets begins with *Atalanta of the East.* Though he had been contemplating a Greek ballet, in the mode of *Daphnis and Chloe,* he decided upon a Greek tale with an Eastern patina superimposed. The result was only partially successful.

The story of Atalanta played by Pearl Argyle would fit comfortably into the diary of a twentieth-century woman marathoner. Atalanta had a will of iron. She refused to marry unless her future husband was victorious in a race with her. Any man who lost, however, would be killed immediately. In spite of this condition, her beauty inspired many young men to contend. Atalanta had the advantage of running fully clothed and armed while the young men were naked. With the aid of Aphrodite, a young man, Milanion (or Hippomenes) triumphed over her. Aphrodite helped Milanion with the gift of three golden apples that she happened to be bringing back from her orchard garden on Cyprus. By rolling these apples off the course, other young men prevented Atalanta three times from outstripping him: for, whether from curiosity, or from greed, or because she wanted him to win, she stopped to pick them up and was thus overtaken. Tudor managed the race with

> extreme cleverness for the apple is dropped every time the two, Atalanta/Sita and Vikraam danced by Hugh Laing pass across the stage. Diana Gould played the Goddess who gives him the apple. Atalanta loses her head at each lap. Tudor put the opening scene of *Atalanta* in the Garden of Hesperides with three goddesses swaying in an exquisite group and yet somehow suggesting the immobility of statues [the Hesperides were nymphs who guarded the golden fruit of an apple tree]; the entrance of Laing begging for the apple to help him outdo Atalanta in the race; and the race itself. [Apparently the race was structured so that the runners entered and left to complete imaginary laps and the spectators rotated to follow their flight.][56]

In a substantive article in the *Sunday Referee,* Constant Lambert quoted Jean Cocteau's statement that "one may derive art from life, but one should not derive art from art."[57] He meant that in *Atalanta* Tudor assimilated poses from Indian dancing and sculpture, which gave his ballet a static look. "Not only is it derivation at second-hand, but it inevitably gives a choppy line to the choreography as in tableaux vivants."[58] Lambert, who originally helped Tudor arrange the music for the ballet, observed that Tudor created his ballet to Debussy's "Pagodas," a series of Javanese airs that apparently "lost all their savour by being played on the piano." An appetite and fascination with exotic music and dances had been encouraged by recent performances of various Indian dancers, notably Uday Shankar. Furthermore, Tudor had been studying with a Javanese dancer who provided a whole new lexicon of Oriental movements, a treasure trove of movements to be integrated into ballet technique. In her early concerts in Europe and later in America, Ruth St. Denis had also explored this innovative synthesis. Nevertheless, Constant Lambert cautioned Western choreographers about their attempt to use Indian motifs. He noted, "Shankar excelled in the combination of mime and dance without sacrificing one to the other whereas Western choreographers (with the exception of Fokine) seemed unable to produce convincing miming that is at the same time plastically significant."[59]

Apparently the ballet also suffered from its slow pace and the fact that its composition needed a more careful structure. But the set designs by William Chappell were highly regarded, as they were fanciful without being frivolous. The colors of the costumes, striking ones, were black, silver, yellow, and gold. The dancers wore bracelets of bells and tinkled; their fluttering hands displayed long gilded fingernails. Photographs of Pearl Argyle showed her with a severe as well as mysterious facial expression. She wears what appears to be an unusual and ornamental headpiece. Rambert called Argyle as Atalanta a great beauty in her Eastern make-up, in which she looked a real Nefertiti, and also recounted the wonderful appearance of Hugh Laing (mahogany-hued) that compensated for the drawbacks. "But after a few performances we dropped that ballet—though I was sorry to lose the few really inspired moments in it and the decor and costumes by William Chappell."[60]

During rehearsals for *Atalanta,* Tudor's relentless focus on the movement qualities of his dances overrode any other concern. In pointing out a major difference between Ashton and Tudor at that time, Mary Clarke retold the story that the dancers in *Atalanta* were mystifyingly situated on the side of the stage while other action in the ballet was taking place. Rambert asked Tudor, "What are those people doing?" Tudor replied, "They are watching a race." Rambert responded, "But how will the audience KNOW they are watching? They have their backs to us. You must think sometimes about the people who are REALLY watching."[61] Clarke elaborated that at the time Ashton and Tudor were fairly friendly and that Ashton was already a more experienced and well-known choreographer:

> For Ashton it was always the audience that mattered in those days. He was anxious to keep things short, witty or charming—never, at all costs, to be boring. His question was always 'What will they think?' 'How will they react?' Tudor, on the other hand, never thought of the audience at all.[62]

In a testament to the creative awareness of Rambert in the early 'thirties, Mary Clarke in *Dancers of Mercury* noted that Balanchine's Les Ballets 1933 and de Basil's Ballets Russes swept Londoners off their feet with a cast of splendidly exciting Russian dancers. "No one was more ecstatic than Rambert. During the empty years, since the death of Diaghilev, she herself had given London more beautiful ballet than any other person."[63] After Diaghilev died, the Russian ballet or Russian renaissance was born again in Monte Carlo in 1932 under the direction of René Blum and Colonel de Basil. On July 4, 1933, it came to London for the first of six visits. "With it began the London era of balletomania."[64]

Tudor was developing a reputation for being a fine actor-dancer and received wonderful notices for his roles that demanded more acting than virtuosic dancing. In addition, on September 28, 1933, he composed and danced a duet with Freda Bamford in the opera *Faust,* also at the Sadler's Wells Theatre. Still filling in as a male dancer for the Vic-Wells Ballet Company, Tudor played Man in *La Création du Monde* October 30, 1933, at the Old Vic, to music by Darius Milhaud. The ballet, by Ninette de Valois, achieved a modest success. He also danced in Frederick Ashton's *Les Ren-*

Antony Tudor in Frederick Ashton's An 1805 Impression, Récamier, *1933.* (Photo by Angus McBean)

dezvous (December 5, 1933). At that performance, a critic chastised the impolite audience for arriving late for the ballet and chattering at the top of their voices during the show. In April 1934, he distinguished himself in the role of the President in the Sadler's Wells production of *The Nutcracker.*

Tudor's interest in the theater was unflagging. In May 1934, he seized the opportunity to create an "interlude" for T. S. Eliot's pageant play, *The Rock.*

The Legend of Dick Whittington (May 28, 1934)

Technically considered a series of poems, *The Rock,* which "reflected Eliot's growing passion for incantation," capitalized on the excitement of the chorus as "it suggested 'the ultra-dramatic' condition of something which utters itself. . . . More than any other resource of the dramatist, it permits speech to be uttered 'beyond character,' somewhere between dramatic speech and music."[65] Eliot's pessimism about art permeated these lines in *The Rock* where Eliot proclaimed that even in the highest forms of music and art, or in the ultimate symbol of the church, man can never hope to capture the divine reality. "Therefore, we thank Thee for our little light, that is dappled with shadow."[66] In addition *The Rock* has a further, perhaps less subtle story to tell; it alludes to the materialistic and lingering poverty of human beings that Eliot sensitively acknowledged, and bemoans man's pathetic, miserable moments of time on earth. "A group of workmen is silhouetted against the dim sky. From farther away, they are answered by voices of the London Unemployed. . . . 'We stand about in open places, / And shiver in unlit rooms. / There shall be one cigarette to two men, / To two women one half pint of bitter ale. / In this land / No man has hired us. Our life is unwelcome, our death / Unmentioned in 'The Times.'"[67] Yet Eliot's message sounds out loud and clear—that the futility of life may be abated with an abiding belief in the church of the Christians.

In contrast to the occasionally eloquent, oratorical poetry of *The Rock,* but in concert with Eliot's idea that men and women often toil without recompense in their daily lives, Tudor created a scenario that he probably read in a children's nursery book or saw in a pantomime. And it tied him to cat-loving Eliot, as one of the chief characters in the dance interlude was a cat. *The Legend of Dick Whittington* is based upon an historical personage who lived during the late fourteenth and early fifteenth centuries. It is the instructive tale of a poor young man who rose to become Lord Mayor of London. He was also a generous benefactor of the city who gave gold to the deprived and neglected. According to the old story, well known to London children, Whittington went to London to work as a kitchen scullion in the house of a merchant, Mr. Fitzwarren. He was so badly beaten by the cook's maid that he ran away, but as he escaped, he heard the bells of London tolling, "Turn again Whittington . . . Lord Mayor of London." With renewed courage, the brave boy returned to the merchant's home, where his master's daughter, Alice Fitzwarren, befriended him. Whittington owned a cat, which he sent as a venture on a ship, and he became a rich man when the cat received enormous sums of money from the King of Barbary for ridding the ships of rats and mice.

Despite the fairytale quality of this legend, it may not all be fable. This is the period of the plague in Europe when cats were a valuable defense against the rodents that spread the dread disease that killed one-third of Europe's population. And so, because of his wonderful cat, Whittington succeeded in becoming Lord Mayor of London. Of course, he married Alice. The story of Dick Whittington was made into a ballad that was printed as a play in 1605, and long afterwards was used as a tract to encourage virtue in young readers.[68]

No doubt this whimsical ballet interlude, with its beloved London characters, offered a welcome relief to an audience a bit groggy from the pessimistic oratory of Eliot's choral exhortations.

Six months later, Tudor and the Rambert dancers presented an ambitious new ballet that was very different from his previous work.

The Planets (October 28, 1934)

The Planets is Tudor's first substantial ballet in which he works with symbolic meanings as well as particular qualities of human nature. Apparently the first ballet made to this music by Gustav Holst was choreographed by the German modern dancer Harald Kreutzberg at the Berlin Staatsoper on May 9, 1931. Fernau Hall declared that Tudor's talents were now fully prepared, his apprenticeship over, and he was ready to tackle a more serious theme. *Cross-Garter'd, Lysistrata,* and *Adam and Eve* were all relatively light works, though based on heavy scripts. In *The Planets,* Tudor composed the parts, not only to reflect the music, but also to create the atmosphere and lyrical movement that reflected different planets' meanings.

When the brilliant artist and scene designer Hugh Stevenson, who also occasionally danced for Rambert, conceived the idea for *The Planets,* he showed the costumes and the scenery to Rambert, and she entrusted the choreography to Antony Tudor.[69] Stevenson provided a different backcloth for each section with varying arrangements of tall, slender pillars against a night sky. Rambert told the story that Tudor tried out his first ideas for the movements in this ballet with Rambert's children on the beach at Dymchurch, and it was after looking at the waves breaking on the shore that he invented the basic movements in "Neptune."[70]

In an interview with Maude Lloyd, she and her companion, Muriel

The Planets—*"Neptune," 1934. Nan Hopkins, Antony Tudor, and Kyra Nijinsky.* (Photo by J. W. Debenham, courtesy of the Rambert Dance Company Archive)

Monkhouse, nicknamed "Tiny," remembered that the ballet was episodic and not a carefully integrated whole. For instance, at the end of the ballet, all the planets do not return.[71] Five years later, when Tudor added the Mercury solo for his newly created London Ballet in January 1939, he probably felt that the "Mercury" section was a stronger ending than the original "Neptune." *The Planets* had one of the largest casts—fifteen dancers—Tudor had ever worked with, for the eighteen-foot Mercury stage could barely accommodate such a large number of people.

Fernau Hall observed in April 1988 that Tudor chose parts from the long musical score's suite of seven movements that he felt would work best in his dance. Since they were not essentially connected, Tudor could pick and choose. Some interesting notes on Holst's musical score of *The Planets* give insight into the ballet. For example, Holst was by no means an astrology aficionado, but he understood the value of prediction. One might presume, although Holst denied this, that his score for "Mars," written shortly before the outbreak of World War I, prophesied the kind of mechanized conflict of tanks, machine guns, and airplanes that was to claim so many lives.

The moods of each section of the music were strongly contrasted, and

Tudor matched them by adopting different choreographic styles. In the first section of *The Planets,* "Venus," two mortals, a boy and girl, are born. This episode was apparently the most like conventional ballet. The planet Venus as well as the movement of the stars make the two white-clad lovers meet and part, then meet again a little closer and part, and so on till the end when they are in each other's arms.[72] Photographs display Maude Lloyd as an exquisite Venus wearing very long black hair and a white, filmy, sleeveless dress. Lloyd recalled that all of the movements in Venus' section were absolutely smooth, without any jarring steps. "There were no jumps; it was all poses, *pas de bourrées,* as if I were floating and weightless." The arm movements were extremely modern as in many of Tudor's ballets, stressing hand gestures and long lines rather than soft and round shapes. She acknowledged that she adored *Planets.* "We did it first of all with three sections: Venus, Mars and Neptune. I was the Planet Venus, with Pearl Argyle and Billy Chappell as the two lovers under my—The Planet's—influence. There were four girls who were Venus' retinue. Tudor gave me very, very slow movements. I remember doing arabesques with strange arm movements. The ballet started with me bourréeing round on the spot with one hand going up and down, very slowly over my head, my head very much to the side, and my body undulating as it turned. What I did was very adagio—big slow movements." One of the "strange" very modern movements that Venus did was one in which the arms were horizontally held above the head (right elbow over left hand) and they crossed each other as the leg moved through from front to back, not unlike one of the movements in *Dark Elegies.*[73]

In apparent contradiction to Maude Lloyd's description of the "Venus" section, John Percival in *Dance Perspectives* asserted, "The Venus danced by Maude revolved about movements that were mainly curvilinear, flowing and rather conventional. The lovers, a couple situated downstage, move with one another until the effect of the planet is felt by them."[74] In an interview, Maude Lloyd demonstrated some of the movements that obviously combined a very expressive hand and arm vocabulary with ballet movement.

The "Mars" scene was choreographed, or "written" as Maude Lloyd said, on Hugh Laing, who was an extremely dynamic mover. The costume for Mars was red with wide trousers and bands of material across Laing's bare chest and around his arms. This section, with much stamping and fist thrusting, had dances in the tradition of Kurt Jooss ("tortured," Tiny recalled). "Mortal and planet leapt, stamped and grovelled in fierce convulsions; everything was extremely contorted and emphatic."[75] The movements were percussive and weighted into the ground. Apparently they were highly modern, aggressive, and angry. The scenario indicated that the Mortal born under that planet is destined always to fight—and in the end destroys himself.

A dancer in the corps, Rosemary Young, wrote up comprehensive notes and drawings of the corps de ballet's movements in the section "Mars."[76] Some of Young's material clearly evokes the spirit of Tudor's choreography: "At the end of the second phrase, the girls turn three times on the spot, clockwise, using both feet and gradually becoming upright, the arms being raised above the head, the hands clenched, they reach this position on the accent, then sink swiftly into original position, which is crouched, the left knee crossing the right, neither on the ground, body forward, head down, upper arms continuing line of shoulders, i.e. parallel to the floor, forearms at perpendicular shape," etc. Other examples of his movements, according to Young's writings, are shuffles, body tilts in knee-bent positions, lunges with the head leaning to the side, arms circling, shuffling backwards, drops on knees, swaying forward keeping body straight, falling straight back so that body is on the ground from head to feet, *pas de chats* (spring right, spring left), landing, legs together, in deep *plié* crouching, arms crossed over chest so that elbows jerk forward and towards each other. This step-by-step description proceeds for thirty-nine pages. Athough it is impossible to know exactly what the ballet looked like, or what the soloists were doing, the notes beautifully suggest a rich and varied movement vocabulary with an openness to modern ideas that was unique at the time.

Tudor composed the third scene, "Neptune," on Kyra Nijinsky, daughter of Vaslav and Romola Nijinsky, with Tudor as a "satellite" planet. In "Neptune," the Mystic, mood was almost everything, aided by the pensive colors of dark brown and green, Holst's introduction of a wordless female chorus, and by his use of very slow movement. Maude Lloyd, who replaced Kyra, remarked:

> The movements were very difficult, because Holst's music for Neptune has no discernible beat. It just goes on . . . and on. And Antony wrote very loosely over the music. Really one had to count in one's head in order to finish with the music. [Though Tudor deplored a dancer who counted, in this case there evidently was no choice.] And he created all those expansive, floating movements—very effective. The Neptune section started off with funny little oriental foot movements, with Antony as the Planet "satellite" behind her. Rather Japanese, its opening, I remember. Then it worked up and Neptune's movements got larger and rounder, ending up with a circle. This Neptune passage featured Tudor who was in the background while Kyra was mystically moving downstage. She is not aware of Tudor who played the planet. You don't know you're being influenced. Some of the movements were very

rounded, smooth, though there were many jumps, very big ones, very difficult, right on the beat.[77]

The "Neptune" section had mystical overtones, as did "Venus."

The fourth section, "Mercury," which was added for the performance of the London Ballet on January 23, 1939, gave the ballet some vivacity and amplitude. The background lighting of gray and blue suggested moonlight. The music contained very complex cross rhythms combined with "a lot of very quick beats" that made it an exciting conclusion. Guy Massey danced Mercury, and Peggy van Praagh played the Mortal.

Rambert praised the balance and structure of this ballet. She pointed out that in each section there was the planet and the mortal born under that planet and satellites to provide more movement. Astronomy was thrown to the wind! Cryptic comments about Tudor's interest in astrology were made by some reviewers, although most thought the ballet a worthy and important piece.

Unpleasant criticisms of the ballet focused on the "Mars" section. The critics wrote that it succeeded poorly, as its tone and look were too much like the Central European dancing of the period, like what might be seen in the Mary Wigman or Harald Kreutzberg concerts. These conservative attitudes resurfaced later in response to *Dark Elegies*. Percival suggested that, at times, the exaggerated tone of the violent and angular dancing, without Laing's hypnotically commanding authority, tended to look comical.[78]

Another "star" performer, Kyra Nijinsky, attracted some attention. Rambert remembered that when Kyra offered to join Rambert, she proudly told Rambert that she had danced her father's role in *Spectre de la Rose*. Rambert gave her Markova's role in Ashton's *Mephisto Valse*. Maude Lloyd added that Kyra resembled her father closely; she had his long neck and heavy legs. Kyra performed with Rambert during the autumn season of 1934. In *Dancers of Mercury*, Mary Clarke recounted that Kyra was studying in London with Nicholas Legat, the great Russian teacher, whom many of the Ballet Club and Vic-Wells dancers went to for extra lessons. "Kyra was delighted to dance for her parents' friend."[79] Arnold Haskell wrote in *The Dancing Times* of March 1935: "She is a dancer who understands, whether instinctively or otherwise, how to use dancing in order to express her emotions. Her technique is not great, but so beautiful in quality is her natural movement and so convincing her mime, that she can divert one's attention from her shortcomings."[80]

The Planets remained in the repertory of the Rambert Company until 1950.[81]

Tudor was not the only one experimenting with innovative dance movement. The London dance world had been shaken and inspired by many new notions. Peggy van Praagh, Agnes de Mille, Hugh Laing, and, of course, Tudor himself emphasized in their recollections of the time the fermenting importance of the various de Basil ballet companies that visited London, and Balanchine's Ballets 1933, the Jooss Ballet, the solo performances of Vincente Escudero and Uday Shankar, and other groups of Kathakali and folk dancers. De Valois and Ashton at the Vic-Wells also absorbed these exotic currents and cross-fertilizations. Apparently the tightly knit Tudor dancers would often follow him to these performances and then return to his studio in Chiswick for long discussions. At the time, Tudor studied with German dancer Anny Boalth, a Central European who taught him the principles that the *Ausdruckstanz* and *Neue Künstlerische Tanz* movements were concerned with.[82] With a diploma from the Laban School, Boalth opened a studio in 1931 in London and continued to teach there, advertisements in English magazines tell us, until the mid-'thirties. Also, Boalth was invited to appear in the first performance of the Camargo Society at the Cambridge Theatre on October 19, 1930, as a token representative of Central European dance. Perhaps Tudor became intrigued with her work as a result of that performance.

It was especially the newly put-together Les Ballets 1933 company that bowled London over, initiating remarkable ballet with the brilliant young George Balanchine. In *Dance to the Piper,* Agnes de Mille described as fresh and hopeful the choreography that characterized the season of the Les Ballets 1933, organized by Edward James with George Balanchine as choreographer and James's wife, Tilly Losch, and a thirteen-year-old Tamara Toumonova. This group was followed by the star-studded appearance of de Basil's Ballets Russes at the Alhambra in 1933 with such impressive dancers as David Lichine, André Eglevsky, Tatiana Riabouchinska, Alexandra Danilova, and Anton Dolin. A number of these dancers worked in the original Diaghilev ballet when Fokine began his radical changes in the ballet vocabulary. Tudor's clique had much food for thought as Tudor proudly admitted that he was a loyal priest of the Fokine cult, which proposed that dance movement must conform to the dramatic and musical content of a ballet.[83] Tudor evidently viewed the symphonic ballets of Léonide Massine with slight mistrust as, unlike Fokine's, Massine's aesthetic values seemed superficially motivated. Marie Rambert's dancers, however, joked that she burned incense every time Massine's name was mentioned. But Tudor did gain insight into his own work by watching Massine's work. Both *Les Présages* (1933), to Tchaikovsky's Fifth Symphony, and *Choreartium* (1933), to Brahm's Fourth Symphony, were important departures from the traditional scenarios of the Ballets Russes. Tudor also absorbed from Massine the possibilities of ensemble dancing that could be explored for a dramatic and even a philosophical purpose. Brooding questions of life and death, destiny and fate composed these symphonic themes.

Yet the Ballets Russes's influence occupied a small part of Tudor's esthetic concerns. Peggy van Praagh stated that in 1933 all of the Rambert dancers visited the performances of the (recently exiled from Germany) Kurt Jooss Ballet. Jooss had won a Paris competition with his ballet *The Green Table* in 1932. As a satiric denunciation of World War I, Jooss's ballet both surprised and moved the London dancers. His social and political themes displayed integrity and originality with costumes that were unembellished and clear in their presentation of character. Without large sums of money for scenery, Jooss experimented with lighting designs in order to do the work of defining locality, mood, etc. Tudor valued Jooss's use of lighting because it concentrated on choreography and created mood. Van Praagh noted that the Jooss dancers were not classically trained ballet dancers, and this explained why pointe work was not emphasized in his works.[84] More important, Jooss developed his dances in the descriptive terms of Laban's and Wigman's expressionistic movement where steps reflect and are impelled by emotional experiences. These intensely psychological notions became a framework for the vocabulary that Tudor explored in the body of his choreography, and a rich language for communicating with his dancers during rehearsals.

At the time that he was choreographing *The Planets,* Tudor took any jobs that offered choreographic possibilities and some spending money. He arranged dances in the opera by Jean Philippe Rameau, *Castor and Pollux,* November 1934, for the Oxford University Opera Club. His dancing did not seem to be improving at any great rate now that he was deeply involved in composition. For example, several years later, in 1937, Fernau Hall criticized Tudor in Andrée Howard's *Rape of the Lock:* "His *grands jetés en tournant* were distinctly elephantine. However one can always rely on Antony Tudor for a good makeup and fine acting,"[85] which seems like a backhanded slap in the face. Literary themes were usually the starting point of Andrée Howard's choreography, and when she heard an inspired reading of Alexander Pope's *The Rape of the Lock* at Susan Salaman's, "she translated it into witty and attractive choreography."[86] On opening night, October 10, 1935, Pearl Argyle danced Belinda, Frank Staff impersonated Sir Plume, Andrée Howard played the sylph Ariel, and Tudor interpreted the role of the Baron who stole a lock of Belinda's hair. Howard, an eminently talented artist and designer, was quickly becoming one of the more distinguished choreographers of the Rambert clan.

After *The Planets,* Tudor again turned to a classical theme for inspiration. His next ballet appeared a little less than six months later. Again it was

The Rape of the Lock, *1935, by Andrée Howard. From left to right: Andrée Howard, Elisabeth Schooling, Frank Staff, Pearl Argyle, Antony Tudor, and Peggy van Praagh.* (Courtesy of the Rambert Dance Company Archive)

something entirely new, as his only habit was the habit of not repeating himself.

The Descent of Hebe (April 7, 1935)

The program reads: "I. Prelude: Hebe, while serving the Gods, trips and spills a cup of precious nectar. Overcome with shame she attempts to steal away, hoping that her carelessness may have passed unnoticed. But Mercury arrives with a message from Jupiter which banishes her to Earth as punishment for her fault. II. (Dirge). Night, with her horsed chariot, awaits Mercury's signal to leave for Earth. He arrives, followed by Hebe, whom he presents to Night with the injunction that she be taken to Earth. Hebe, however, refuses to enter the chariot. Thereupon, Night conjures up a vision of Hercules whom Hebe will meet on Earth. Hebe is captivated and, now as eager to depart as formerly she was reluctant to leave, leaps into the chariot. III. (Pastoral). Hebe arrives on Earth; she dances with children when she encounters Hercules who woos her. IV. Fugue, Apotheosis."

51

The Descent of Hebe, *1935. From left to right: Elisabeth Schooling, Hugh Laing, and Maude Lloyd.* (Courtesy of the Rambert Dance Company Archive)

There is a note at the bottom of one of the programs that the musical installation for *The Descent of Hebe* is by "His Master's Voice" (an RCA record). Antony Tudor stage-managed this production.

Though less ambitious than *The Planets, The Descent of Hebe* was well received. The ballet appeared on the same program as Frederick Ashton's *La Valse chez Madame Récamier;* Pearl Argyle was Madame Récamier, with Antony Tudor as her suitor. Frank Staff appeared in the *pas de quatre.* Also in *Le Cricket, a Sporting Sketch,* by Susan Salaman, Tudor played the Bowler and Frank Staff the Umpire. Tudor seems not to have been in Ashton's *The Lady of Shalott,* but Hugh Laing played Sir Lancelot. The last piece on the program was Ninette de Valois's most successful, *La Bar aux Folies-Bergère,* a ballet. Maude Lloyd danced the dramatic role of La Goulue. William Chappell made the sets and costumes for *Shalott* and *Bar aux Folies.*

The story of how Tudor chose the myth of Hebe and Hercules is of some interest. In 1933, Marie Rambert's husband, Ashley Dukes, presented a play by J. V. Turner, produced by Rupert Doone, entitled *Jupiter Translated.* In the middle of this play there was a short suite of dances, in which Hercules and Hebe were the principal characters. Beginning with these two characters, Tudor evolved the scenario for this ballet, which is set to Bloch's

Concerto Grosso. The Nadia Benois costumes remained in the Mercury wardrobe, and when Tudor decided to make a ballet about Hebe, Nadia Benois designed some additional ones and some new settings. "The designs were not only 'ravishing,' but remarkably ingenious. The black horses of Night's chariot leaped across the backcloth, and the synchronisation of the dancers' movement with the progress of lines of pink clouds above the stage gave a wonderful illusion of flight."[87]

Traditionally, Hebe is the goddess of youth, the daughter of Zeus and Hera. She has the power of restoring the aged to youth and beauty, and sometimes she appears as cupbearer to the gods. The myth tells that since death would not come to him, Hercules would order his own demise. He commanded a great pyre to be built on Mount Oeta, and when Hercules reached it he knew he could die, and he was glad. "This is rest, this is the end." As they lifted him to the pyre, he lay down on it as one who at a banquet table lies on his couch. He asked his youthful follower, Philoctetes, to hold the torch to set the wood on fire; and he gave him his bow and arrows, which were to be far-famed in the young man's hands at Troy. Then the flames rushed up and Hercules was seen no more on earth. He was taken to heaven, where he was reconciled to Hera and married her daughter Hebe.

Perhaps the most exciting aspect of this production was its visual beauty, especially in light of the stage's limited size. As lovely as the scenery might have been for the spectator, though, the decor posed huge obstacles for the dancers. Maude Lloyd reflected, "Tudor had two rows of cut-out clouds on the stage which came down when the scene changed to heaven. We had to dance in between them and jump over them, and not hit them with our feet—Rambert was absolutely adamant! We had curtains which acted as wings on each side; they were dark blue on one side and gold on the other, and they just swung around if you wanted a change. Of course, when you jumped out onto the stage, there was very little space, but if we touched those curtains and made them swing, we got into real trouble so that doing 'Hebe' was a real nightmare!"[88]

Retrieved from the Rambert Archives is a photo of Tudor carrying Elisabeth Schooling (who replaced Pearl Argyle as Hebe) in a lift. The following notation was made on the photo margin:

> The clouds which hide Antony's feet are made of wood painted pink and gauze. They are hoisted into the flys when the scene changes to earth, and once they got stuck half-way up during the black-out. The lights went up and revealed them dangling in mid-air.

Janet Leeper, a witness to the early Tudor ballets and author of *The English Ballet,* exclaimed:

Who will ever forget those dances to Bloch's "Concerto Grosso for Strings and Piano," the prancing horses of the goddess painted on the backcloth by Nadia Benois, the aerial car—strictly static—giving the impression of heavenly ascent . . . amid fantastic semicircular clouds of pink gauze disposed about the tiny stage? It was a flight of fancy beautifully realized, the foretaste of much to come.[89]

Maude Lloyd reminisced about her role as Night: "I was Night, and I had a chariot. This slow movement also exploited endless variations of arabesques for me. I had to unhook the reins in the dark and climb up on my chariot and stand waiting. I was so giddy in the dark. Agonized. Then I had to come down and dance between these clouds. We all did. Sometimes there were about six of us dancing between these terrible clouds. They were made of hard wood. Two wires held the whole lot of them. I had to do a double pirouette in between the two; I always felt I was going to overbalance."[90]

Maude Lloyd mentioned that soon after the premiere, Elisabeth Schooling played Hebe, and Peggy von Praagh took over her role as Night. Schooling described Tudor's unique way of choreographing: "I think Tudor was distinct from other choreographers of his day in the manner in which he worked. He seemed to draw his inspiration from the centre of his chest, standing very still and emitting small gasps, sometimes rousing Hugh to anger!"[91] Agnes de Mille quoted Lloyd as saying that the rehearsals for *Hebe* were "two and three hours of agony for a single posture or jump."[92]

Maude Lloyd delighted in remembering a very special *pas de deux,* a "fabulous duet which occurred between Hebe and Hercules, that is, Argyle and Tudor." Gradually Tudor discovered his powers as Hercules with gay, leaping movements that recalled episodes of pastoral scenes and Greek gods from anacreontic ballets fashionable at the beginning of the nineteenth century.

When questioned about *The Descent of Hebe,* Fernau Hall remembered Tudor's comment that

the most difficult genre of all in ballet is the abstract, that is, in abstract ballet, the images have to be that much clearer. Antony's Fugue in "Hebe," the second half of the ballet, relies upon the music to drive the dancers on, and it does. The Fugue section displayed Tudor's discerning use of the dancers in play with the driving force of the music. As the fugue theme repeated and built, Tudor manipulated the dancers on that tiny stage in such a way that the dance reached a crescendo unmatched in his other ballets.

Hall also asserted that Nadia Benois was exceedingly sensitive to the music and "she responded to the stratified nature of the Fugue by laying down horizontal strips of clouds that were one foot high. During the fugue scene the dancers entered from the sides, different voices of the music coming and going in Fugue form. In other words, Benois visualized the Fugue on the stage."[93]

Contemporary spectators also suggested that a marvelous moment in *Hebe* occurred during the Fugue when Hugh Laing and Pearl Argyle danced a duet. But some of the technique that Tudor presented to Laing proved too difficult for his abilities at that time. After all, Laing had started to dance very late by ballet standards:

> After a few rehearsals with Tudor, Hugh decided to ask Miss Margaret Craske if she would help him to strengthen his *batterie* or quick beats so that he could easily execute the fast steps. She gave him private lessons, starting from the very beginning like a child. Within a few weeks he had improved enormously and was prepared for the duet with Pearl.[94]

Until this moment, Tudor had played character roles such as Malvolio, the Mystic in *The Planets,* and one of the husbands in *Lysistrata.* In other words, he had not taken on the heroic qualities of grand gestures and skillful technique. As Hercules, he was teased: "And Antony Tudor, who had designed the ballet, had the stature and the grace of Hercules. But only a mythical perruquier could have produced so 'permanent' a wave in his beard."[95]

In a 1986 interview with Hugh Laing conducted by Marilyn Hunt in New York, Hugh told her that she would have liked *The Descent of Hebe* very much:

> But it couldn't be done here as it could only be done on a small stage because of the staging. The clouds had to go up. We had to jump over the clouds when we came down. And when we were coming down, I was Mercury and I was taking Hebe and I had to go from this thing down to my knee in arabesque. And the clouds were going up at the same time. So it looked as if I were going through the clouds and Hebe was hanging on behind me like this with the clouds moving. It was very beautiful to look at. But it couldn't be done on a large stage because you could never have arranged the cloths and scrims and wires.[96]

Despite technical problems, Cyril Beaumont praised the Ballet Club for refining the genre of the *ballet intime* when he wrote, "The Ballet Club has evolved a repertory of ballets in which resource, economy of expression and imagination are so strongly developed as to form an independent trend of ballet. These *ballets intimes* call for an added responsiveness from the audience, as any work that relies for its effects on suggestion rather than statement must necessarily do." Tudor's talent was at ease in this minimalized form and, later on, he found it difficult to translate this ability to a larger corps and a greater stage space.

But Beaumont's review continued harshly: "Choreographically the work is disappointing. Its incident, so cleverly seized in the decor by Nadia Benois, peters out into a number of spineless passages on the stage, and the dancing patterns tend to be repetitive rather than progressive. Imaginatively, the movement is never equal to the decor and Hebe's banishment becomes a wan little tale."[97]

In the same vein, Arnold Haskell pointed to a defect in Tudor's creative process that he grappled with all of his life:

> The *Descent of Hebe,* Antony Tudor's ballet, IS STILL UNFINISHED!!! More adequate rehearsal shows the necessity of several changes in both choreography and setting; the ballet is, in fact, planned on such a scale that it will stay unrealised until it reaches a larger stage . . . its last movement, a Fugue, works ten dancers into as satisfying a contrapuntal intricacy as I know; and, as a whole, it is a strangely mature work, in which a lovely formal structure is no longer overlaid and blurred by overmuch detail (as it was in *Lysistrata*) and in which an exactly right and light use of postures from the Grecian urns promises a true synthesis, a style which will in no way dispossess the qualities of the essential Ballet style, but mesh into its achieved texture new material—for one thing, giving the upper part of the body more significance than it has had.[98]

At a repeat performance, *Hebe* was given a fairly good critique: "The invention falters a little in the Pastorale but the first two movements are fanciful, and the final Fugue builds up strongly into a massive ensemble for the whole company."[99]

Maude Lloyd referred to a paragraph in Rayner Heppenstall's book *Apology for Dancing,* in which he predicts that this new, young choreographer, Antony Tudor, should be watched. Though he had not as yet had the opportunity to create works for a large theater, "his work as a whole seems to me richer, and more pregnant than any other work at present being done in ballet. *The Descent of Hebe* produced at the Mercury last year and not yet offered to the press, has not realized itself fully until it reaches a larger stage" (which, Maude added, it never did). "It is, I am convinced, the most vigorous piece of choreography that London has seen since Balanchine's *Cotillon.*[100] Lionel Bradley's *Journal* spoke of *The Descent of Hebe:*

> The Ballet Club phonograph records were played on a gramophone apparently located behind the stage and the piano part was played on a piano with the lid down in front of the stage. It wasn't bad, but it would have been better if the gramophone had been nearer and one was never aware of the turn over

of a record. The costumes and sets were good. The pastoral back cloth [he describes all three] was like a child's impression of the country with sheep, deer and birds on a landscape of field and trees. In the Fugue, a little incongruously, the horses were on the back cloth without the chariot. Night and her attendants were in black gauze material . . . Not all the dancing and whirling was really significant but it was all pleasing and the principals did well. Antony Tudor looked more muscular than expected. And I was glad to see Hugh Laing in a full part.[101]

Schooling cautioned:

This ballet was always a hazard because of the two rows of clouds through which we danced in the first and last scenes. Going up and down in blackouts things were fairly wayward and worked manually. If all went well, it was a delight to dance the final fugue. If not, they might remain at chin height leaving the choice of going under, or over at your peril, or they might even descend unexpectedly, probably at speed during the Fugue.[102]

The Descent of Hebe remained in the Rambert repertory until 1950.

Tudor continued making opera dances through 1937. He arranged some for Rossini's *La Cenerentola* (May 1, 1935) at Covent Garden, with Alicia Markova.[103] His relationship with the Oxford University Opera Club also endured. A news clipping from an Oxford newspaper mentioned that he and Maude Lloyd had to dance without their costumes in *Lac des Cygnes,* as their costumes had been stolen. Tudor went on to create the following opera divertissements for the Sadler's Wells Opera Company—*Schwanda the Bagpiper* (June 3, 1935) by Jaromir Weinberger; *Carmen* (June 4, 1935); and *Koanga* (September 23, 1935) by Frederick Delius.

Looking back upon this formative period, it is difficult to know about Tudor as a younger person growing up in London, his inner thoughts, fears, and ambitions. His school years and those just afterwards must have been terrible for him. As he was a relatively shy, "unaccomplished" (as he said), and extremely sensitive young man, one wonders at the endurance he needed in order to keep working at the Smithfield Market, doing what his family and those around him expected. In a way, it was both inevitable and miraculous that he broke away from that working environment to Rambert's home and studio, living with her, Ashley Dukes, and their two children in a new world of emotion, books, plays, poetry, and ballet above all! He used to peruse her library and borrow books from her all the time. It was with Rambert that he reshaped his persona, changed his name, his accent, and became what he truly wanted to be, an artist. He was his own finest creation.

In the early years of his choreography, Tudor charted new ways of presenting highly charged dramatic moments for his small-scale ballets, or *ballets intimes*. *Cross-Garter'd*, *Lysistrata*, *Adam and Eve*, and *Atalanta* all brought clearly drawn characters into wry, comic relief. Their sophisticated humor and charming ironies appealed to cultivated English audiences. In the tradition of Fokine, each ballet devised movement that suited its subject. With *Atalanta*, Tudor played with an exotic dance vocabulary to fit his balmy idea of a Greek myth in Oriental trappings. His lessons in Javanese dance helped him fuse ballet with Oriental gesture. In his serious work, *The Planets*, as Rosemary Young's notes confirm, Tudor experimented with a more modern movement vocabulary, using shifting levels, unorthodox arm positions, tilting balances, and many torso movements. In *The Planets* and *the Descent of Hebe*, Maude Lloyd and Fernau Hall verified that Tudor developed his interesting musical approach wherein the specific beat of the music or the melodic line was not necessarily the primary focus of his choreographic phrase. Rather, he moved his dancers while sustaining the motivation and the tension that his characters expressed.

The following years brought Tudor into an historical position that sealed his greatness and perhaps made it difficult for him ever to reachieve this limelight. With *Jardin aux Lilas* and *Dark Elegies*, Tudor discovered his métier. Until this moment, Tudor had been testing his mettle.

CHAPTER THREE

Revelation of a Major Talent

*B*y 1935, Tudor was beginning to be appreciated for his outstanding and innovative choreography and was at last commanding his own following. Several important dance viewers and critics, such as Rayner Heppenstall, Cyril Beaumont, Lionel Bradley, and to a lesser extent Arnold Haskell, admired him, as did the dancers in Rambert's company, who labored ceaselessly to satisfy him.

The reasons for the dancers' attraction to Tudor's classes were many. Maude Lloyd often spoke of his intellectual passions, especially as a teacher. He brought with him to classes and rehearsals at the Ballet Club an already mature knowledge of music, art, and literature. He was bookish. He adored the stage, any form of it—music hall, opera, revues, musical theater, drama. The influence of those interests is very evident in his choreography: his calculated use of strong literary themes, his open attitude (later) toward the new art of television, and his numerous choreographic forays into opera and musical shows. These enthusiasms helped him overcome weaknesses in his classical ballet technique, so that he brought with him to the classroom and the studio a breadth and understanding of movement that fascinated his dancers. He also created for himself character roles in his ballets that only he could carry off, with a style and concentration that galvanized and consolidated whatever was happening onstage.

In the early 'thirties, Tudor began work on an idea that, like *Dark Elegies,* harked back to Nijinsky's ritualistic *Sacre du Printemps.* It concerned the ancient Finnish epic *Kalevala.* Tudor rehearsed with several Rambert dancers, including Maude Lloyd, Peggy van Praagh, and Kyra Nijinsky. Hugh Stevenson's remarkable designs still exist.[1] To the modern eye they look distinctly stylized, Nordic, and heavy. Tudor developed a full-length scenario for *Kalevala,* and in the summer of 1936 he went to visit Jan Sibelius in Finland to discuss the Magyar origins of the Finnish people. Sibelius had composed, in 1892, a successful score called *Kalevala,* but still he politely declined to collaborate with Tudor. Although the October 1935 *Dancing Times* advertised the coming performance of the "Kalevala Epic," it was never produced, and Tudor often regretted not having been able to realize the piece. Instead, he created *Jardin aux Lilas,* also based on a story by a Finnish author, which represented a major breakthrough in choreography.

Jardin aux Lilas / Lilac Garden (January 26, 1936)

With the premiere of *Jardin aux Lilas,* Tudor discovered the sweeping, lyrical, and pure classical technique he became famous for. This was Tudor's first attempt at a style that was later to become characteristic of his work—the ballet of situation, of atmosphere, of nostalgia. As Hugh Laing put it, "I think that *Jardin* was the first time that Tudor began using relationships of people for what they call 'psychological ballets.'"[2] For this masterpiece, Rambert paid Tudor the sum of seven pounds.

Jardin aux Lilas, *1936. From left to right: Hugh Laing, Maude Lloyd, Antony Tudor, and Peggy van Praagh.* (From the Performing Arts Research Library)

Tudor's discerning interest in theater artists of the period included a taste for playwrights whose characters were driven mainly by psychological concerns. Among the literary influences on the ballet was Eugene O'Neill's *Strange Interlude,* a play in the mode of stream-of-consciousness, whose heroine has married unhappily. In addition, Tudor's penchant for French literature, specifically Guy de Maupassant, Marcel Proust, and J. K. Huysmans, heightened his sensitivity to environment, to smell, to the importance of place and location. Tudor's original idea had been to make a ballet based on a short story by the Finnish author Aino Kallas. (Fernau Hall has said that a similar story by the French writer Georges Ohnet was also a source.)

> The Kallas theme concerned a young peasant couple, about to marry, and a landowner who proposes to exercise the antiquated *droit du seigneur.* Separately, Hugh Stevenson had the idea of a ballet, romantic in mood, set in a lilac garden. The designs for costumes and backdrop were startling. The ideas fused; the *droit du seigneur* was rejected as being impracticable (perhaps too dated?) and was replaced by a forced marriage.[3]

Jardin aux Lilas epitomizes the ballet of mood and ambiance. In fact, according to Laing, on opening night, "quantities of lilac water in atomizers were sprayed throughout the theatre."[4] In this subtly passionate work, the ardent feelings of several members of English high society are revealed and questioned. The scenario centers on a marriage of convenience; probably one of the earliest efforts along balletic lines to indicate a concern for contemporary human values.

The scene represents an evening party in a lilac garden, and the period is Edwardian.[5] Agnes de Mille astutely observed:

> It was Antony Tudor who first put his women in long dresses, the Edwardian dresses his mother wore. The effect was startling and a real shock to the imagination. The audience was called upon to accept the balletic gesture as a form of simple dramatic communication. It was also asked to watch women who looked like their mothers and aunts kicking over their heads, or wrapping their legs around men's bodies.[6]

The curtain opens on Caroline, a bride-to-be dressed in a white-lace strapless gown, in the garden with her fiancé, who wears a dress coat. He is the Man She Must Marry. The light of a full moon brightens the garden with a fearful glow.[7] The couple are polite but stiff and bored. Caroline stands with her back to her fiancé. Suddenly, her former lover enters, and she is greatly disturbed, warning the Lover not to approach. She walks away with her fiancé mechanically, in processional steps. They go off, and guests come into the garden and dance. They twirl, sweep in and out of intimate moments, and interrupt other lovers without knowing how jarring their intrusions are. Caroline returns and meets the girl who was her fiancé's mistress (An Episode in His Past), but she ignores her. Finally, alone with her lover, she anxiously holds him; they seem to dare destiny with lunging walks forward, watching to see that they are not observed. When they leave, the serious, upright fiancé enters with his former mistress. Thus, the whole ballet, according to Lionel Bradley, "consists of meetings and partings thereby presenting an extraordinarily successful idea of emotional progression."[8]

The fiancé and his former mistress dance; the situation has been reversed. But their duet is laden with guilt, memories of stolen pleasure, and anger. As the music gathers force and volume, the mistress repeatedly jumps into his arms, throwing herself at him, although he does not welcome these abrupt attacks. These are symbolic movements, especially when "she runs across the stage and leaps at him, to be caught horizontally at shoulder height. Both literally and metaphorically she is throwing herself at his head."[9] Stephanie Jordan insightfully commented that "whenever the mistress appears, the music is 'animato' or 'allegro.' Even when she first enters with the other guests, a succession of descending eighth note patterns moves the music out of the static and hints at future tension. She is desperate, but certainly not resigned."[10] But the fiancé, simmering and withdrawn, visibly tries not to remember her.

Guests pass and repass, and the tangled situation becomes increasingly tense. There are touching solos for Caroline and her lover. Each paces back and forth, reaching for what could be his or her heart's desire. At one moment, Caroline is alone onstage. She begins to pirouette swiftly; her lover comes in just in time to catch her as she finishes spinning and save her from falling. Bradley said that when they danced together, "the cross-currents of feeling, the haphazard but skillfully organized meetings and partings, the revelation of character and the general sense of frustration all combine as evidence of Tudor's imagination and skill."[11]

Three young women guests, who seem to share Caroline's secret, dance with the Lover. They point to a distant mark in the sky as he joins them. They are reaching for a falling star, which signifies the intense spark of a love quickly extinguished, whereupon the four dancers then clasp their hands in prayer. It is a moment, Maude Lloyd asserts, that has not been given enough weight in recent productions.

The lovers do not wish to sacrifice each other, yet the music and the dancing accelerate toward an undeniable end when the fiancé must claim Caroline and marry her. Caroline and her lover appear, running across the back of the garden. When they re-enter, they find the fiancé and his mistress in the garden. The fiancé turns away from her and suddenly catches Caroline about the waist to lift her high above him. Caroline's arm points straight up over her head; she resists holding him. Other guests enter, and the fiancé bends down to kiss her hand while she falls over in a lunge, her body arched and hanging away from him, as if in a faint. Then occurs a moment of true invention when all the elements of Tudor's dance—music, plot, and a driving movement style—seem to be resolved. Tudor silences the stage with a tableau that warns us of the inevitable.

During this striking tableau, the music sounds a "brutally victorious theme for full orchestra fortissimo for a few bars as everyone poses motionless; the act culminates, the past is relinquished."[12] The fiancé walks away, and Caroline kisses her lover for the last time; he hands her a branch of lilacs to take with her. Her fiancé returns with her cloak, puts it over her shoulders, and presses her arm to her side as she makes a despairing gesture to her lover. The ballet finishes as the couple walk off with sad dignity—leaving the lover alone on the stage with his back to the audience.

Tudor felt a particular affinity for the Edwardian period. Although Edwardians questioned established institutions, they knew enough not to disturb their affluent status quo. Edwardian prosperity and glitter, social stability, and spacious ease represented halcyon times before the cataclysm of World War I. Thus Caroline's forced marriage for money and social position represented a perpetuation of ritualized and convenient class choices.

Tudor stressed that the necessity of "keeping up appearances" is central to the scenario in *Jardin:*

> It's important to me that a woman of this period always remains a lady, no matter what. And the expression of this elegance and sensibility means that the torso must be a part of the movement. The dancers today don't use their torsos. They don't express things with their bodies as much as we did and this is very necessary in a ballet like *Jardin,* where the smallest movement has to make an effect. . . . In consequence, the dancers/characters maintain an erect, almost stiff stance in public. It is only fleetingly, in private, that they can express their real feelings. Truth and honesty become furtive abstractions.[13]

According to Lloyd, "Madame Rambert used to say Tudor had a devil inside him which he had to keep tightly held back, and every now and then it burst. In the movements of *Jardin,* when the lovers meet each other, they should be burning with love, but you didn't show your emotions when other people were around. So all those little moments when they meet are the only moments in which they show their true feelings."[14]

Tudor instinctively understood the disguises people wear in order to separate their feelings from a certain persona. The Tudor who grew up as William Cook in the East End watching the activities of West Enders realized that behind the froth and wealth lay real feelings and conflicts. As Jack Anderson wrote:

> *Jardin* shows pain behind splendid trappings. . . . Tudor suggests depth of feeling by such small details as the placement of a hand in a social gesture and by such large-scale devices as the way the Mistress literally and figuratively throws herself at the man she loves, and by the movement, in the most dramatic pause in all ballet, in which time momentarily stands still for Caroline.[15]

Peggy von Praagh wrote:

> When we were working on *Lilac Garden,* [Tudor] instructed us, "You all know what the Edwardian period is! These people all live in a time when social conventions were strong; now go and read some Edwardian novels. Try and get the background of your character: you can find it in any Edwardian novel." I remember that he gave me a little piece of paper on which he had written the particular qualities he wanted and told me to go away and think about them.[16]

Lloyd remarked how different the qualities of the English are today from those of Edwarian times:

> It is almost impossible for our dancers to understand this restraint. One of the reasons why I think the young find his ballets difficult to interpret is that you have to convey that you are feeling emotion without showing it. Whenever anyone else is on stage, it was the tiniest gesture, this drawing back you had to do. And you know, it's the strangest thing, but you can show an emotion by such a tiny movement. You can feel people's muscles tensing on stage even if you can't see them.[17]

One had to feel the constraint of the Edwardian period:

> You can't make a big gesture. It has to be small but it has to be effective, and that means you have to have tension. And that ballet is full of tensions and fluid movements all mixed up one after the other. It's quite an extraordinary ballet in which sometimes you have to show your emotions with your back to the person you're emoting about, or standing side by side without looking at them. You still have to let the audience know what you are feeling. This is not an easy thing unless you're feeling it yourself.[18]

Finding the music for *Jardin aux Lilas* was also complicated. Clive Barnes disclosed:

Rambert and Tudor went, as was their custom, to the HMV record shop showrooms in Oxford Street in search of suitable music. They started by listening to Fauré, but decided, in Rambert's words, that he was "melancholy more than sad." Then they came to Chausson's Poème for Violin and Orchestra, and the choice was made.[19]

Tudor recalled:

> With *Lilac Garden,* I had the subject resting in the back of my head. I needed music expressing early twentieth-century romantic, Victorian. In Japan, they call it Meiji. They think there that *Lilac Garden* is a very Japanese situation ballet. Looking for music to fit the perfume of lilacs, the emotions of four principal characters . . . a piece that could not be too long. . . . I began with Fauré's Ballade. It didn't work. I finally turned to Chausson.[20]

According to Elizabeth Sawyer, Tudor originally chose Fauré's *Ballade* and worked with it for some time. He subsequently decided to use Chausson because the Fauré did not satisfy the intensity of his scenario, although the score by Chausson was "rather exaggerated and sentimental." In addition, Sawyer pointed out, the *Poème* is one of the most predictably rhythmic pieces in an era that was known for predictable rhythms.[21]

Chausson composed it in 1896, and its expressive mood suggests a typically romantic theme, a long-drawn sigh for all things irrecoverable and unobtainable. Noel Goodwin has suggested that, the next time one views the ballet, one should "observe how the rise and fall of the musical outline colours the successive choreographic episodes, while at the same time the phrasing of the dance is paced independently of the music."[22] Laing recalled that, while Tudor spent many hours "digging," he was not wasting his time:

> Maude and myself and Tudor would work night after night with our accompanist, Norah P. Stevenson, at the piano. And [Rambert] was not sympathetic to *Jardin.* . . . She wanted to be in control of everything. It was her theatre and she wanted to know what was being done and how it was being done. Well, there are certain times when you cannot do things other people's way.[23]

After a while Tudor forbade Rambert to attend rehearsals of *Jardin.*

Tudor's discordant manner of choreographing is legendary. Even in his early years in London, he struggled to discover *le mot juste,* just the exact line and movement. Lloyd said that Tudor always created his *pas de deux* on himself while partnering her. If he could not necessarily see the moves, he felt them:

> He knew from inside himself what he wanted to say and he left it to the audience to see whether they liked it or not. He wrote the ballet in passages and he often threw out what he did after a night of thinking about it. He would

65

say, "I don't like what I did last night—too easy, too flowing—it's not what I want." In England there was a very small band of absolutely devoted admirers. I don't want to sound snobbish about this but they were all people of intellect. . . .[24]

The response to *Jardin aux Lilas* confirmed, with some exceptions, Tudor's growing reputation as a remarkable choreographer. Rayner Heppenstall alluded to the fact that there remained some rough edges to polish away, but *Jardin aux Lilas* is the first true *ballet intime,* the first occasion of pure choreographic lyricism."[25]

In Arnold Haskell's review, Hugh Stevenson received high praise for his decors while Tudor's choreography is described as "interesting in patches, but it is too jerky as a whole. At times one feels that the choreographer has been frightened of appearing sentimental so that he has deliberately gone contrary to the suavity of the music."[26] Haskell also mentioned the novelty of using a gramophone at the same time that the company had "the rare privilege of Yehudi Menuhin as accompanist."[27] This practice was changed, according to Haskell, to the more usual live accompaniment. Haskell predicted that the ballet, which he likened to an "evocation of a painting by Tissot, a French garden of the 1870s," would become "a period picture of great charm and distinction." Strangely enough, Haskell proclaimed that Tudor's portrayal of the fiancé was the best performance.[28]

Several years later, at the opening of Tudor's London Ballet at Toynbee Hall, Lionel Bradley suggested that the greater width of the stage took away something from the urgency of the meetings and partings:

> In the general dance, when the quartet of principals is doubled by the addition of two pairs of friends the effect which on a small stage seemed circular, now rather resembled the four corners of a square. And I was sorry that they omitted the presentation of a bunch of lilacs to the bride by her lover just before the end. The bride now has a spray of lilacs fastened to her skirt.[29]

Jardin aux Lilas initiated a new trend in ballet. Laing said:

> I think that *Jardin* was quite a new approach to ballet. This was before [Ashton] had done things like *Foyer de la Danse.* It wasn't real people in what could be interpreted as a de Maupassant story. And this quite upset [Rambert]. For one thing it wasn't passionate enough. People weren't tearing their hair and falling on their knees and all that nonsense. They were just behaving like rather refined people of the upper, or lower-upper middle class and not in any way fairy princes or princesses. If you will look at *Lilac Garden,* they are all classical steps there. . . . There is not any modern movement. It's all your glissades, your arabesques, your pirouettes, your attitudes, everything. It's all there.[30]

The role of Caroline was first danced by Maude Lloyd—and she is still coaching young Carolines. One critic praised her "noble serenity and deep expressiveness"; another found her "clarity itself, in line and with at the same time, the utmost subtlety and refinement in the qualities of movement." Ironically, Tudor had had another dancer in mind for the role—Pearl Argyle. He was intrigued by the beautiful ballerina he had used in his *Atalanta* and *Hebe.* Originally, the ballet was to have been a triangle among Argyle, Peggy van Praagh, and Lloyd. But Argyle was dancing with the Vic-Wells ballet and did not have enough time to learn the role, so that van Praagh, who had been rehearsing the second lead, became An Episode in His Past and Lloyd took over the lead.

Very soon after the opening, Tudor eliminated the identities of the people at the wedding party; as of February 9, 1936, they were called simply "Guests." When Tudor mounted the ballet in New York for Ballet Theatre, he increased the number of Guests to eight, and the revised production has remained the basis for subsequent stagings. The Ballet Theatre 1940 production, called *Lilac Garden,* had designs by Raymond Sovey after the originals by Hugh Stevenson. The setting by Tom Lingwood, originally for the Royal Danish Ballet, was later adopted by American Ballet Theatre. These designs have been used for most later productions. *Lilac Garden* remains to this day a centerpiece in the repertory of American Ballet Theatre.

The critical and financial success of *Jardin aux Lilas* pulled Tudor farther away from Rambert. No doubt she knew that Tudor was moving in his own direction and could no longer be controlled, but as the matriarch she needed to have her say in all aspects of her company's activities. It is also not unlikely that Rambert was jealous of Tudor's creative potential.[31]

Jardin aux Lilas was restaged by Tudor at the New York City Ballet on November 30, 1951. Many regional and national companies have subsequently performed *Jardin,* including the National Ballet of Canada from 1953 to 1956, the Royal Ballet in 1968, the Paris Opera Ballet in 1985, and the Kirov Ballet in Russia in 1991.

The atmosphere in the 'thirties that fostered Tudor's increasingly personal approach to movement belonged to a new London, a place that encouraged émigrés and foreigners to pursue their careers. London was becoming an important artistic crossroads that had significant influence in the creation of western art. Just as a number of artists had fled Russia to come to London after the Russian Revolution, so many artists were fleeing from Hitler's Germany, and their works hinted at the frightening political turbulence. While the other arts were in ferment with new ideas, modes of ballet-making tended to be conservative. One of Tudor's achievements was to open doors of the ballet studios to the new concepts that were in the air, thus changing the choreographer's approach to the ballet vocabulary, structure, drama, and characters.

One of the strongest representations of this new approach came almost exactly a year after the premiere of *Jardin aux Lilas,* when Ballet Rambert presented the masterpiece *Dark Elegies.* It was the culmination of a number of years of intense study and work for Tudor.[32] He brought to his passion for ballet a knowledge of composers and an interest in the writers Sigmund Freud, Henri Bergson, and Marcel Proust, among others, who influenced artistic and philosophic thought during the 1930s. He had already attained recognition for *Lysistrata* (1932), *The Planets* (1934), *The Descent of Hebe* (1935), and of course *Jardin aux Lilas* (1936). *Dark Elegies* was the one that moved Tudor from artistic backwaters into the mainstream of Western European culture.

Dark Elegies was austere, sparse, and finely etched. Its treatment of so touching a subject as the death of children defied conventional ballet theatrics. Tudor instead introduced a new movement vocabulary for a ballet that looked very much like modern dance. He was not alone in exploring philosophical and challenging subjects of a serious nature; despite geographical distances, his sympathies lay with gifted American and German dance choreographers such as Mary Wigman, Kurt Jooss, Hanya Holm, Martha Graham, and Doris Humphrey.

Dark Elegies (February 19, 1937)

When *Dark Elegies* opened in London at the Duchess Theatre, Tudor's enthusiastic following had to pause for reflection. The Mercury Theatre, where Ballet Rambert usually performed, was not well suited to the larger corps of dancers in *Dark Elegies* and its extensive time on stage.[33] With *Jardin,* Tudor had proved his abilities as a choreographer with a rare understanding of the classical technique. Now it was time for a larger stage and a broader canvas for his brush.

Tudor achieved the representation of exaltation through suffering in *Dark Elegies* in several ways, the most important of which was his use of Mahler's *Kindertotenlieder.* Tudor probably discovered this music in an early-1930s recorded interpretation by Hermann Rehkemper.[34] The music for *Kindertotenlieder,* or "Songs on the Death of Children," was composed in 1902 to five of Friedrich Rückert's eponymous poems. Two of Rückert's children had died of scarlet fever, and these poems are elegies for his loss. In

Dark Elegies, *1940, Ballet Theatre. From left to right: Hugh Laing, Lucia Chase, Antony Tudor, Miriam Golden, and Jerome Robbins.* (From the Performing Arts Research Library)

1900, Mahler began work on setting the poems to music. Sadly, he and his wife Alma also lost one of their daughters to scarlet fever in 1904. Tudor was inspired to use the song poems, but in a very subtle way, led by the sense and rhythm of the words. In the original production of *Dark Elegies,* a singer intoned the songs, accompanied by two pianos. The songs, Rückert's poetry, described the sad event that befell the seaside community.[35] All the dancers as well as the singer wore the simplest of Everyman costumes, drab shirts, pants, dresses, and shawls. The women wore scarves over their hair. A general gloom hovered over the whole ballet—an environment of harsh weather and little daylight—to which the choreography described a particular agonized response by each solo dancer. Thus the universe of the ballet is meant to revolve, not only around the death of the children, but also around our shifting perception of time of day. Listening to the singer, the audience is gradually drawn into the tales of each of the couples; they become, with the singer, sublime witnesses to the disconsolate accounts.

First Song

(*Langsam und schwermutig, nicht schleppend*/Slowly and painfully, not dragging)

Now will the sun arise so bright
As though no death had come last night.
On me alone the blow did fall;
The sun shines equally on all.
You must not hide in night unmending,
But let it fade in light unending!
(shaken convulsively:)
One candle snuffed out in my tent;
Hail the world's joylight heavensent!

Tudor justly begins his cycle with women moving ritualistically through their seemingly predestined patterns. The stage appears dimmed by a cloudy rim. Six women kneel in a circle, perhaps already in prayer. Another woman enters, her hand reaching very high above her head, the other hand holding her elbow. Tudor uses a minimal vocabulary of sways, walks, *pliés,* and kneeling while the hands rest in the lap. There is a quiet wailing in the body, as there is in the music, with occasional gestures or arm movements that have no meaning *per se,* but seem to tell us about the women's dolorous experiences.

These gestures often recur. In one of them, the hands seem to float above waves, and then scoop down, telling how the sea has taken the children, as the women circle around each other. They reach into the distance, and sometimes contract their torsos forward as if sobbing. When they rise on pointe, it is as if they have not, or are simply trying to reach up looking for where the children have gone. The music swells, or opens up, and the solo woman leaps, as if pained by the other spaces she has been in. Tudor takes his time, builds slowly, and asks the viewers to join the group as they watch the women grasp one another or move mechanically, holding their elbows close to their bodies in isolated grief.

Second Song

(*Ruhig, nicht schleppend*/Calmly, not dragging)

Now well I see, why often you would throw me
Such glances always full of strange dark fire,
O eyes! . . . as if within a single moment
One look would concentrate your entire power.
I never guessed, because the fog enwrapped me,
All raveled by remorseless blinding Fates,
That this lone beacon came but as a summons
For your return to light's primeval homeland.

And with your shining, this you'd try to tell me:
If we could we'd gladly stay beside you.
But this is something Fate has not allowed you,
So look on us, for parted soon we shall be.
These days what now are only eyes to you
In future nights you'll only see as stars.

Tudor chooses a man and a woman for this song, one of the most touching *pas de deux* in his repertoire. A single woman rushes in with a man following quickly after her; he lifts her straight up into the air as if to offer her in prayer. As in many of Tudor's passionate duets, the lifts are breathtaking and comment clearly on the couple's experience together. She falls backward, swooning in a circle. He lifts her, both her knees up and high, head back; then she folds over, her torso and head touching her knees.

Many of the lifts have a lyrical stop-point and then switch to another shape in midair. One of Tudor's signatures is the lift that ends with the woman kneeling on the floor. The couple seems to be pleading, seeking answers, she in a *pietà* position couched on his leg, his hand extended below his heart, framing and as if holding it. Their movements emphasize swaying, vacillation, tipping back, falling forward; round and round. They seek each other's hands as they move round each other. One of their last lifts is a stunning fish-dive which he throws her into, her face skimming the ground as he holds her waist. Two men enter and approach him as he kneels by her prostrate body. They lift her above their heads, a parting gift to the higher spaces.

Third Song

(*Schwer, dumpf*/Painfully, muffled)

When your Mommy dear steps beside me here,
And I turn my head, and see her near my bed,
Not upon her face falls my anxious gaze,
But upon a spot, *there,* below her heart,
Where your little face would turn to meet
Mine so joyfully, with her, my little girl,
  Just as you used to do.
When your Mommy dear steps beside me here,
As her candle gleams, to me it always seems
Just as though you were skittering after her,
As you used to do! You, your daddy's image,
Ah! how quickly my light of joy winked out.

I never guessed, because the fog enwrapped me,
All raveled by remorseless blinding Fates,

That this lone beacon came but as a summons
For your return to light's primeval homeland.

In the Third Song a young man is featured as he moves through a group of six dancers. Restrained and tense, the group of dancers hold together in chains, their arms around each other. They flow in patterns like folk dancers, weaving in and out, crossing each other; everyone seems to know what the next change is. A young man emerges, framed by a group of six, then eight others. He seems to be commenting to his lost child on the sadness that his wife is suffering. His movements, at times jerky, with sharp kicks and slices of the hands, become rigid, his foot inadvertently shaking underneath him like a strange tic he cannot control. The group resumes its patterns.

Fourth Song

(*Ruhig bewegt, ohne zu eilen. Schlicht, aber warm*/Restful movement, not hurriedly. Alla breve. Weakly but warmly)

> I often think, they've only gone out walking.
> Soon toward their home again they'll be returning.
> The day is fine! O, don't be afraid—
> They're only on a long holiday!
> O yes, indeed, they've only gone out walking,
> And really soon back home they'll be returning.
> Don't be afraid, the day is fine:
> They're only climbing up to yonder heights.
> They've only pulled ahead of us while walking,
> And didn't want to be so soon returning.
> We'll catch up with them there on yonder heights:
> In bright sunshine . . . the day is fine!

The Fourth Song concerns another woman of the group. The lyrics suggest a dramatic sense of isolation, of frenzy as if her children had been lost and must be found. For a while, a woman from the group pretends the children are off on a stroll and will return. The light of the sun pours through the hushed atmosphere; the children have flown away to "yonder heights." The bereaved will follow their flight. Tudor begins to press the dancers with the music, as if trying to shake them out of their numbed sorrow. The solitary woman moves among the eight dancers who are sitting in a circle in groups of two and three. She darts from one place to another, almost jumping on pointe from group to group, in ever-quickening desperation. She seems driven by a force beyond herself. Alone in the center of the circle, her arms cradle a shadowy being. The men approach her, sheltering her with their arms. At the beginning of the song the stage light is bright, but gradually it darkens, and people close in from all sides of the stage.

Fifth Song

(*Mit ruhelos schmerzvollen Ausdruck*/With restless anguished expression)

> In this rough weather, in all this tumult,
> Never would I have sent my children away!
> Someone must have dragged them out—
> I could say no word to stop it!
> In this rough weather, in all this storm,
> Never would I have let my children away!
> I was afraid that they'd get sick . . .
> That is now an idle thought.
> In this rough weather, in all this horror,
> Never would I have sent my children away!
> I worried lest they die next morning . . .
> Now is not the time for worry.
> In this rough weather, in all this horror,
> Never would I have let my children away!
> Someone must have dragged them out—
> I could say no word to stop it!
> (Slowly, like a lullaby)
> In this rough weather, in all this storm,
> They rest as though in their mother's home.
> Not panicked by any tempest,
> By God's own hand they're covered.

In the Fifth Song, the scene begins with the darkness and flashes of light. A single man starts to move, while the others are leaping and flying in a circle; they move like a herd, anxious but obedient, as if driven by some monstrous force. He seems caught in their swirls and crisscrosses. Then he finds the center of the turbulence, turns, and falls to the floor; he gets up and falls again. His turns, *tours en l'air,* and allegro, are effortless, but his movements achieve a psychic heaviness and emotional passion that none of the other songs had attained. The pace quickens, the rhythms push on. The space seems to get crowded with people thrashing in different directions, returning at times to their comforting, primeval circle as darkness falls.

The fifth song continues with Scene 2. Resignation: Light returns to the stage as if the sun has broken through the clouds. The mood has changed: a quiet falls on the group; the men stand while the women kneel. There are now twelve people together (one sees how the whole ballet has built, even in numbers). Four women face each other and move toward each other on their knees, suppliants and stoic celebrants. They stand and return to the floor patterns of earlier songs. The couples take hands tenderly and depart in a procession. The solo woman from the first song is left behind to follow, repeating the grief stricken movements of her entrance. They have danced their ceremony and achieved a sense of equanimity.

Tudor's simple and economical choice of movement, a restricted but highly refined vocabulary, imparted to the ballet an individuality and self-reflection that virtuoso technique could not have communicated. Following the principles of Rudolf von Laban, Tudor sought to expand the space in which ballet arms and legs normally flowed. By including thrusts, jabs, and tensions of the fingers, hands, legs, and feet, Tudor discovered what might be termed "emotional space." His pedestrian dancers, with their odd stiffness like English Morris dancers, suggest an improvisational style, shifting the way square dancers do as they discover new patterns. The willed spareness and restraint of the movement stands in contrast to the florid emotionalism of the music.

Occasionally, and powerfully, the dancers pause, and nothing happens. They freeze into suggestive inaction. This use of stasis heightens other moments when the dancers either fold into delicate and complex groupings or explode into a frenetic and charged use of space. These controlled spatial patterns constantly remind the audience of the personal and repressed inner turmoil that is ready to burst out, even destructively, if unleashed. It is perhaps vengeance against war.

Peggy van Praagh believed that in the case of *Dark Elegies* Tudor responded to the mounting destruction of the Spanish civil war. She mentioned in an interview with Margaret Dale that "some people think that *Elegies* was Tudor's comment on the raid on Guernica in Spain where so many children were killed." The first dancers in *Elegies* represent the women who lost their children: their despair, their agony. She goes on to say that "by the end of the ballet, the time of the storm, everyone gets very excited. Tudor resolves this fury in the last song when a complete change occurs. This is acceptance, resignation. Tudor always told us that the last song was almost a kind of ecstasy that comes after tragedy."[36] In Tudor's words, "The ultimate importance is that the audience be reached and held in the mood of the work, that the audience dwell in that mood during its performance. . . ."[37]

The critics in London responded to *Dark Elegies* as if they had received a good shaking-up. They heard about the work; they respected Tudor's creative genius.[38] Fernau Hall wrote, "The dances in *Dark Elegies* are as revolutionary as Massine's *Les Présages*."[39] A. V. Coton appreciated the treatment of the subject "for its refreshing absence of any balletic or dramatic clichés of situation, gesture or dress normally associated with the idea of dying. *Dark Elegies* was the first and remains the greatest English work of ballet on a serious theme conceived and executed throughout with a wholly novel technical medium and with a refreshing freedom from the sometimes embarrasing conventions of what is proper to the depiction of death in the theatre."[40]

But several critics found *Dark Elegies* morose and somber. Tangue Lean felt that the music was very "slow and unvarying." Lean also noted that "a symphonic ballet of any length is likely to suffer from steps that are either irrelevant or agonisingly personal."[41]

Other critics such as Mr. Forsyth, dumbfounded by the music and the untraditional choreography, focused on Nadia Benois's compelling backdrops: "The best thing I can say about it is that Mme. Benois, who is a well-known painter, was its saving grace."[42] Similarly, Cyril Beaumont downplayed the movements and praised the decor: "A florid dancing idiom has no place in this work; neither has the free expression of grief in the manner of the Central European choreography. . . . All moves somberly towards a calm epilogue superbly captured in the most distinguished decor this company has shown."[43]

By now, Tudor had developed his trademark method of composition, bringing his dancers into the process mainly through readings, but also through suggestive and improvisational methods. For him, steps had to have motivation. He sculpted the dancers' characters in a Stanislavskian way and introduced an intimacy to their understanding by questioning them and teaching them how to listen to the music with emotional breadth. Thus, with meticulous care, Tudor was able to bring individuals within the community to life.

Van Praagh related a compelling account of rehearsing with Tudor for *Dark Elegies*. Apparently, either because she was more available to Tudor or because she did not talk back, she provided him with a lot of time for his movement ideas:

> In rehearsal Tudor would work with the one person he had chosen to perfect the style. The fourth dance was originally performed by Agnes de Mille (and was later danced by Andrée Howard most beautifully). Agnes had an unusual quality, but Tudor did not work the dance out on her; he prepared it on me, although I had my own dance as well. He used to keep me back to work on Agnes's dance because she was very busy with her own recitals. I also think he didn't find she inspired him; she was rather critical ("What's all that about?" she'd ask) and I didn't ask any questions![44]

Similarly, de Mille recounted the painfully long time Tudor employed in finishing his dances. "The final heartbreaking moment of *Dark Elegies* was composed onstage the night of the opening with the company in full make-up and dressing gowns. . . ."45

Van Praagh confirmed this story, but expanded it:

> When it came to the dress rehearsal, he had not choreographed Hugh [Laing's] solo. Hugh was a marvelous dancer, but he was not a very quick "study." So the day of the first night, the solo was still not complete, though Antony and Hugh had been up half the night setting it. On the day of the opening we had a stage call at about five o'clock to learn what we had to do during Hugh's solo, because we all acted as corps for each other's solos. Hugh, of course, could not remember a thing of what had been set for his solo, but Antony told us what we had to do. . . . When *Elegies* began things went very well up to Agnes's solo, the fourth song, and then came Hugh's solo. He did the first step and then couldn't remember a thing. He ran round the circle muttering, "I can't remember! What shall I do?" and . . . we all whispered further ideas. The next morning Marie Rambert called a rehearsal. The reviews were dreadful.46

The ballet did catch on with the public after a few performances, and the sense of tragedy was felt by cast and audience alike. During the war, it was tremendously moving, van Praagh said, when "we performed *Elegies* with the London Ballet. In performance, during the last song—with its acceptance of tragedy—we made little bows to each other, there would be hushed silence in the auditorium and you'd suddenly hear someone quietly weeping in the theatre. It was shattering."47

Dark Elegies continues to be performed nationally and internationally. The original Rambert Ballet version was slightly changed when Tudor first restaged it in New York, and he tinkered with it ever after. At this writing the Rambert Company still performs its own interpretation, which in 1989 caused a ruckus, as American dancers who reconstruct Tudor's ballets felt the Rambert rendition of the work was distant from what Tudor had indicated for American Ballet Theatre.

Many companies have performed and continue to perform *Dark Elegies,* including the National Ballet of Canada in 1955, the Royal Swedish Ballet in 1961 and 1963, the Dutch National Ballet in 1972, the Bat-Dor in Tel Aviv in 1976, and the Royal Ballet in London in 1980. Though some critics find the piece stilted and old-fashioned, a rather moving and wonderful television performance staged by Sallie Wilson with the American Ballet Theatre was broadcast in 1990 and demonstrated the continuation of *Elegies'* power to touch its audiences. *Dark Elegies* is in the permanent repertory of the American Ballet Theatre.

In looking back over the entire body of Tudor's work, *Dark Elegies* in 1937 is balanced by *Echoing of Trumpets* in 1963, more than twenty years later. Both ballets emphasize the emotional despair of women bereft; both ballets refer to folk movement idioms; and both find their roots in the outrageous tyranny of German militarism.[48] We now can see that Tudor, like Picasso, was a child of a cruel century reacting in the language he knew best to its wars, its destruction, and its persecution of women and children.

Alone in its time as it seemed, *Dark Elegies* was not without antecedent. Until 1937, Nijinsky's *Le Sacre du Printemps* was in a class by itself. Of all Tudor's associates, only Rambert was likely to have seen *Sacre,* but the parallels between it and *Dark Elegies* are evident. Both attempt to decipher deep communal responses to death, with their communities sharing over-flowing emotional reactions to experiences that cannot be understood on a rational level. The groups draw everyone into a magical circle that then whirls too quickly, becomes too powerful, throwing individuals outward from the center. The formal repetition of ceremonial movements achieves a certain desired intensity in both ballets. Each participant in these rituals is set apart from the real world, and is given a feeling of power from having performed in the ritual, although Tudor's movement choices for the corps are far less differentiated from the ballet technique than Nijinsky's. There is also a musical association. Van Praagh noted that Tudor's inspired relation-ship to Mahler's score breathed a special life into *Dark Elegies,* just as Ni-jinsky in his attention to Stravinsky's score evoked a "primeval relationship of man to earth, man to the elements in an extraordinary way. . . . In both of these ballets, it seemed as if classical ballet had bridged time to join with the beginnings of dancing. . . . That is why they stand alone."[49]

Indeed, these were epic ballets; they distilled the essence of a commu-nity and took it to mythological heights. If toe shoes were not entirely discarded, they were masked by folk movement—flat-footed, turned-in, and parallel. Apparently, in *Dark Elegies,* the solo for Walter Gore was inspired by a national dance that Tudor had seen when he was vacationing in Yugo-slavia.[50] Constantly varying patterns of lines and circles made by defined groups characterize both ballets. Tudor tried to give the impression of a more natural way of moving in order to create real people in real situations, and variations on folk movement satisfied his scenario as it had Nijinsky's. One can also trace its lineage to Bronislava Nijinska. In both her *Les Noces* and *Dark Elegies,* "individuals emerge only to be absorbed again by the group."[51]

CHAPTER FOUR

Tudor on His Own

*A*t the same time that Tudor was becoming a master choreographer and teacher, he fought through the anger that he harbored in Rambert's presence. Though he respected and appreciated her, he had to escape from her controlling, autocratic nature. He arrived in her studio in 1928 and left almost a decade later. After having created nine ballets specifically for the Ballet Club, Tudor broke with Rambert. The same spirit that led him to leave a safe job at Smithfield Market persuaded him to leave the mentor who had recognized his potential and nurtured it. He had tired of arguing with "Mim" and of having to defend his choreographic ideas. De Mille noted that Rambert had no clear idea of the welter of crossed emotions that rocked and paralyzed a heart like Tudor's. Tudor is quoted as saying that "'Mim' had a malevolent disposition at rehearsals that was only assuaged by the artists reproducing everything she expected."[1] After *Dark Elegies* in 1937, he was ready to be on his own.

Within two weeks, he was busy working on some novel ideas for a television show, *Fugue for Four Cameras,* which caused a mild sensation for its inventive qualities. After that, he took every television job that was offered, creating small ballets for BBC-TV nearly once a month for some time. The BBC began experimental transmissions in 1929, and in 1936 officially inaugurated regular programs, which ended only in 1939 as a result of the war. The number of television sets in use grew from several hundred in 1936 to approximately 20,000 in 1939. Conditions were difficult: working on the spot, rehearsing several intense hours, not knowing what would really succeed on camera, never seeing the product repeated.[2] His major producer, Dallas Bower, has described the typical fare:

> I revived on TV a light entertainment form known as intimate revue. This was popular in the London Theatre throughout the 1920s and early 1930s. . . .

Revues always contained ballet in one form or another. . . . I always found Tudor easy to work with, although a forceful disciplinarian in as much as he demanded intense concentration upon the work in hand.[3]

Maude Lloyd commented:

There were two very advanced and rather intellectual television directors at that time—one was Stephen Thomas and the other, Dallas Bower. Both were great admirers of Tudor and they used to get us up there, doing pas de deux and little ballets very often. We became the king and queen of television for a few years. We got ten pounds [approximately $50.00] for one performance and fifteen when we repeated it in the evening because there were no videos unfortunately at that time.[4]

In addition to television work, Tudor also began to give lectures at Morley College and Toynbee Hall. The May 1937 *Dancing Times* announced that Antony Tudor would give a course of six lectures on the ballet for working men and women—(1) The Development of Classical Ballet Dancing; (2) Period Dancing and the Influence of Costume; (3) Ballet: The Synthesis of the Arts, Ballet Composition; (4) Ballet in Rehearsal, Audience in Performance: The Completion of a Ballet; (5) The Use, Misuse and Abuse of Music; (6) Choreographer and Dancer, Ballet, Finite and Infinite—a very sophisticated approach to dance and dance history.

Offering such lecture-demonstrations at Toynbee Hall and at Morley College also helped him consolidate plans for his own company. Peggy van Praagh mentions that Tudor taught classes in ballet history and theory: "Tudor made us demonstrate a class, and explain how a dancer was trained, and how a ballet company operated."[5] Later on, these lectures would be given in exchange for use of Toynbee Hall Theatre.

In several of his lectures, Tudor used the dancers who performed the last Fugue in *Descent of Hebe*. First he would play the Bloch music for that scene, then have the dancers dance without music; finally he would put the music and the dancers together. It was a great success. Another of his approaches was to ask the audience to write down various stipulations for a ballet—whether they wanted to see a *pas de deux* or a solo dance, the time of day, and a simple plot. Then he would gather his dancers and develop a piece, accompanied by gramophone records he had brought with him. He confessed that these little exercises were exhausting, but he was quite convinced of their value.

That spring of 1937, Agnes de Mille and Antony Tudor decided to pool their talents and meager funds to create a company they called the Dance Theatre. Leaving Rambert was not a huge loss for de Mille. She was not a

79

member of the company and, in fact, felt that Rambert hated her. Prior to leaving, de Mille had taken advantage of Rambert's Ballet Club by renting studio space and having her costumes made in their shop. When, at Tudor's insistence, de Mille was given a starring role in *Dark Elegies,* Rambert was so rude about her performance that de Mille packed her bags and left. "Then I suggested to Tudor that we combine our forces."[6] Remembering those days, de Mille said she was astonished by Tudor's often diabolical humor and brilliance. She felt at home in Tudor's somewhat seedy apartment, where he made wonderful meals for her. She mentioned that when she first met Tudor and Hugh it was like a thunderbolt. Their complementary personalities and interests created a fascinating, if somewhat explosive partnership. Hugh Laing was extremely high-spirited, quick-tempered, and emotional, with a serious talent for the art of scenery and costume design, while Tudor "was slow, gentle, diffident, humorous, courteous and much abused. He taught and composed for almost nothing, watching everything with remembering eyes and drank his tea quietly wrapped in his dreams of world ambition. He was a kind of hibernating carnivore."[7]

Tudor and de Mille's new, and as it turned out, short-lived, Dance Theatre scheduled a series of performances at the Oxford University Playhouse beginning June 14, 1937. Unfortunately the Oxford students were in the flurry of final exams, and the houses were practically empty. They closed precisely one week later.

For the Oxford appearance, Tudor and de Mille revived some of Tudor's better ballets, and he created a new work, *Gallant Assembly*. The programs were given variety by de Mille's solo dances. Maude Lloyd could not go to Oxford, so Peggy Van Praagh took her role in the *Descent of Hebe*. She tells the story of how she had to stand on a chair in the wings like an airborne apparition. When she leaned over, no one was there to catch her. "Laing finally arrived and they were convulsed with laughter. Tudor was furious that they had dared to laugh and ruin his ballet."[8]

Several articles at the time described the remarkable artistic qualities that both de Mille and Tudor shared. "The Dance Theatre isn't just another ballet company. They have a real idea. They want to add more American dances and some American dancers." The author went on to say that it would take a year to get the company into its proper shape: "The whole point is to get away from the classical ballet formula and create something new."[9] De Mille's press coverage emphasized her originality and her authentic theater background. As a solo dancer she was compared to Ruth Draper. In her choreographic character studies, "she showed every emotion from bewilderment to sheer funk. The little company has an excellent corps de ballet, contrasting so strongly with what we usually see in the English provinces—flopping clods of young women who look like gentlewomen waitresses refusing a tip."[10]

Yet while de Mille experimented in her character roles with folk dance and other dance styles, Tudor evolved his choreography from a classical foundation. The result of this partnership, according to critics, was a "lack of cohesion in the programmes. Miss de Mille's work remains fiercely individualistic while all but one of Tudor's ballets are in the romantic tradition."[11] A. V. Coton, one of the finest dance reviewers of his time, later proposed that "the outbursts of vituperation against Tudor for the creation of *The Judgment of Paris* [1938] could better have been directed against *Gallant Assembly,* whose initial performance a female summed up succinctly: '*Gallant Assembly?* It ought to be called just Dirty Party.'"[12]

Gallant Assembly (June 14, 1937)

Gallant Assembly was a beautifully produced eighteenth-century Gallic romp. Alive with a colorful setting and bouffant costumes borrowed from the eighteenth-century costume designer Boquet at the Paris Opéra, the action focuses on the illicit skirmishes of bored aristocrats seeking scandalous diversions. Originally conceived by Hugh Stevenson in a stately bronze-green and black setting, with magnificent baroque costumes, four ladies, two gentlemen, and two hired performers mince their steps in this comedy of manners.[13]

The setting for *Gallant Assembly* is a rather solemn formal garden. The men wear *tonnelets,* an eighteenth-century version of the tutu—and nodding plumes, while the women are in elaborately hooped skirts, fancy fortresses for fancy ladies. In a pseudo-pastoral scene, the lascivious activities of a group of "aristocrats in love" are interrupted by a dulcet *pas de deux* by hired performers, who are carried on in a majestic palanquin borne by two Indians worthy of the splendor of a baroque court. The simplicity and innocent youth of the hired couple intrigues, indeed attracts, the "aristocrats in love," who attempt to seduce them but receive dignified rebuffs, which rapidly breaks up the celebration.

According to Lionel Bradley, "there is no conclusive story, it all seems to resolve into the pursuit of woman by man and vice-versa."[14] There did seem to be a fascinating central figure, however, that of an older woman, described by Lawrence Gowing as

> an ineffectual old maid seeking some gallant's regard. She moves in short rapid steps making with her hands movements of rotation most subtly indicative of her state of mind. Her fingers move with the busyness of work at a

Gallant Assembly, *1937 (as performed by Dance Theatre at Oxford, 1937). From left to right: Charlotte Bidmead, Hugh Stevenson, Agnes de Mille, Peggy van Praagh, Antony Tudor, and Victoria Fenn.* (Photo by Angus McBean, courtesy of the Rambert Dance Company Archive)

spinning wheel or needles. De Mille's brilliant, pointed characterization provided the old nymphomaniac with a rapt obliviousness and a habitually self-sufficient air. When she does manage to find a man, like a spider ready with cunning desire, she tries to win him, but as he already has a beautiful woman's passionate interest, the old maid retreats to her web and continues to rotate her uncontrollable hands.[15]

Lionel Bradley added that de Mille's partnering work was "varied by odd shaking movements which at once call to mind a cock and hen."[16] Tudor's appreciation of historical dance led him to reproduce and combine the older baroque forms of dance with contemporary ballet technique.

Despite the good reviews, the show folded precisely one week later. Out of money, Dance Theatre had no more engagements.

Through the summer of 1937 and well into 1938, Tudor maintained his heavy schedule of television commitments and continued to give classes at his relatively new studio in British Grove. He was called the "Jesus" of British Grove because of his rather ascetic bearing and tastes, and dancers flocked to his classes for enlightenment. But his partnership with de Mille had by no means ended. They worked together consistently in classes, and finally, she asked him to choreograph some dances for the show that she had arranged as an adjunct to Nikolai Gogol's *Marriage* at the Westminster Theatre.

Seven Intimate Dances (June 15, 1938)

> Premiere June 15, 1938—"Nightly from 15th June until the 2nd of July.
> *Hunting Scene* Music J. C. Bach, Choreography by Tudor.
> *Joie de Vivre* Music Jacques Offenbach, Johann Strauss, Choreography by Tudor.
> Premiere June 15, 1938, Westminster Theatre
> Cast. Agnes de Mille, Antony Tudor, Hugh Laing, Charlotte Bidmead, Therese Langfield.

Given as a curtain-raiser to Gogol's *Marriage.* Tudor choreographed *The Hunting Scene,* and *Joie de Vivre,* while the other dances were composed by de Mille.

> First Series:
> *May Day*—Beethoven; *The Hunting Scene*—J. C. Bach; *Stagefright*—Delibes; *Pavane* from the film *Romeo & Juliet*; *Ouled Nail*—Kurdish tunes; *Joie de Vivre*—Offenbach, Weston, Strauss; *The Parvenues*—Waldteufel, Strauss.
> Second Series:
> *Burgomasters Branle, Joie de Vivre, Audition, Hymn, The Hunting Scene, The Parvenues* (only six dances noted here).
> Third Series:
> *Pavane* from "*Romeo & Juliet*": *Stagefright after Degas, Joie de Vivre, The Parvenues, The Hunting Scene, Hymn, Judgment of Paris*—Kurt Weill.

Horace Horsnell reviewed the show, "The play was preceded by Miss Agnes de Mille and an excellent little company in a series of character and comedy dances that were delightful in themselves and delightfully done."[17]

Judgment of Paris (June 1938)

Lionel L. Bradley and others give the date as June 15, which was the opening night of Gogol's *Marriage;* but the program for that night in the Enthoven

Judgment of Paris, *staged by Ballet Theatre, 1955. Lucia Chase, Agnes de Mille, and Viola Essen.* (Photo by Fred Fehl)

Collection shows the curtain-raiser as *Seven Intimate Dances.* The program was to be changed weekly, and *Judgment of Paris* must have been given later in the run of the play. Bradley's diary makes it clear that he did not see *Judgment* until it was revived by the London Ballet at Toynbee Hall.[18]

In the most acerbic, witty, and cutting mood, Tudor whipped up what may be termed a "Depression" *Judgment of Paris* as a takeoff on the antique myth, without the original's idealized setting or innocent mood. He might also have been playing a little competitive game with Frederick Ashton, whose more traditional version of *Judgment of Paris* had opened at the Sadler's Wells Theatre one month earlier and was still running.[19]

Although Tudor's penchant for using apples as major props was notable in *Adam and Eve* and *Atalanta of the East,* the apple associated with the mythological Judgment of Paris plays no part in Tudor's eponymous ballet. Rather, he chose to limit almost all references to godlike Olympian characters by shifting his setting to a cheap *boîte de nuit.* Hugh Laing, the multitalented dancer, dreamt up floozy costumes and decaying nightclub artifacts that underscored Tudor's amusing exposure of sordid music-hall dancing.

Perhaps it was the music above all that tempered this strange piece of dance theater and endowed it with melancholy and pessimistic decadence. Kurt Weill's music had already achieved some renown during the famous Les Ballets 1933 season in London, when Balanchine had choreographed Weill's haunting *The Seven Deadly Sins*. David Drew, a musicologist, pointed out that 1933 was the year that Weill was driven out of Germany and his music banned by the Nazis.[20] Drew felt that the two-piano reduction made for Tudor by John Cooke detracted from the initial power of the *Dreigroschenoper* orchestral music. The six numbers excerpted from the original score were *Moritat* (two verses leading to introduction only of Polly's *lied*); *Kanonen-Song; Eifersuchtsduett, Barbara-Song* (starting at the refrain "Ja da Kann mann"); *Seeräuberjenny;* Introduction to Polly's Song; Finale (first *Dreigroschenfinal* [Act 1], last sixteen bars only). When the ballet was revived in New York for Ballet Theatre, it was performed with orchestra.

The scene opens on a poorly lit dive; an air of deadly boredom hangs over the appalling place. Lounging at a table are two bedraggled female entertainers, one of them reading a newspaper; at the second table is another broken dancer and a waiter who exudes an air of purpose and solidity. Suddenly there is a flurry of activity as a customer enters. "He has an air of slightly naïve opulence about him."[21] Upon seeing their potential prey, the lady dancers in their old high heels and fishnet stockings discover a certain amount of inspiration and like wound-up toys begin to vamp back and forth, executing old chorus line dance steps. The customer gazes at these three wrecks preparing to show their stuff and asks the waiter for a bottle of wine. It seems that the customer, a latter-day Paris, will be the judge of a surreal beauty contest, although unlike Paris, he is a bit tipsy and has no interest in any of these scary women.

The first candidate, Juno, presents herself for the customer's attentions. She seems to be younger, more innocent than the others, more anxious to please. Swishing her black lace fan, she wiggles and lunges, waddles and swings her hips. She manages an aura of sultriness but strikes out with the client. When old Venus pops up in her ancient blonde wig, she manipulates three hoops and tries to lure the customer to jump through one of them. Venus' blue satin outfit that reveals her ample charms also has an enormous bow on the behind which she flaunts as she pushes her behind closer to him. The hoops are gingerly moved from leg, to leg, to arms. Finally she indiscreetly places her foot on the table where he is seated and succeeds in capturing his interest; he noddingly gropes for one of her hoops. She proceeds to spin one in each hand near her head and steps into them, triumphantly pulling them over her body. The customer takes another drink and she sits down. The last "goddess," Minerva, struts on with her feather boa, which has definitely seen better days. She tries to breath some life into it as she blows on the feathers. With creaking knees, Minerva tapdances her way through several bars of music. She holds out one end of the feather boa, hoping he will reach for it. Instead, his head is about to hit the table. She vulgarly spreads her legs in a second position, then coyly runs, first away and then toward the sloshed client. He falls back with his head on his chest. Defeated, Minerva and her boa slouch back to the other women. With a momentary flash of life, the client points to Venus, as if to say "Come to me, you've won." They clink glasses and he collapses for good, nothing but a "soggy carcass." The waiter and all three buzzards descend on him for his wallet, gold chain, watch, and so forth. We know that the customer is stripped clean and will never see those café dancers again.

In a 1989 interview, Agnes de Mille, who created the role of Venus, testily remarked:

> Nobody's really done that hoop dance but me. There's the least amount of movement in that dance! Every gesture is a satire of some other kind of bad dancing and I knew what Antony was satirizing. I became Duncan, or I became some other dance artist. With each one, there was a bad odor. And Antony's performance was superb, drunker and drunker and eyes glazing with just a touch of lust that faded as he lost his senses. He was a damn good actor. They can't do *Judgment* any longer. They aren't trained, they can't act, they don't know how to do comedy. They don't know how to look, and listen and wait. That's the real business! But it's another form that's no longer being taught to dancers. Sad!

Dismaying to some and pleasing to others, the show received mixed reviews. Lionel Bradley agreeably appreciated "this sordid satire with an acrid undercurrent of tragedy." In another commentary, he remarked that there was a "bitter tang to its humour, like a salted almond, but the sense of sardonic satire prevents its realism from being merely unpleasant."[22] A critic in the *Daily Telegraph* complained that "Renewed acquaintance with *Judgment of Paris* increases my astonishment that a theme so degraded and so sinister should be sponsored by a philanthropic and educational institution [Toynbee Hall] with the Archbishop of Canterbury as Chairman—but maybe this is an old-fashioned view."[23] This "gem of the gutter," at moments unbearably tragic, still appeals to late twentieth-century audiences. Perhaps the performances lack the sharpness and decadent quality of the earlier cameos, but the meaning remains clear and the message powerful.

Judgment of Paris has received a number of restagings, including those by the Ballet Guild in Toynbee Hall in 1942, the Norwegian Ballet in 1958 and 1961, North Texas State University in Denton in 1962, and the National Ballet of Canada in 1962.

In the summer and fall of 1938, Tudor was hard at work keeping his dancers together, hoping to establish his own dance company. Miraculously, a place, a little money, and the devoted band of dancers came together, and the London Ballet was formed. One week before the premiere at Toynbee Hall, a new and charming "dancey" (vivacious and lyrical) ballet by Tudor was presented as part of a benefit performance by the Imperial Society of Teachers of Dancing for the Swiss Cottage School for the Blind.

Soirée Musicale (November 26, 1938)

The Imperial Society of Teachers of Dancing, which paid for the costumes at Cyril Beaumont's suggestion, was an organization that supervised the quality of dance teaching in England. Tudor himself had several certificates from them, and these permitted him to teach certain grades of ballet and character dance. He explained that *Soirée* was not created as a ballet: "It was done as a demonstration piece for the Cecchetti Society [Imperial Society of Dancing] for an annual meeting."[24]

A newspaper headline proclaimed, "All star dancing with 500 performers. Children too! Ballroom dancing with special praise that should be given to *Soirée Musicale.*"[25]

The charm of the piece was exactly that it did not pretend to be a fancy ballet with extravagant Regency decor (although in 1951, Hugh Stevenson did design a beautiful backdrop as well as new costumes in a production for the London Theatre Ballet). The setting had a blue backcloth and several elegant chairs. A group of dancers entertained one another with "ballet-cum-national idioms in costume suggested by early nineteenth century prints."[26] Tudor balletized various national dances, to the tuneful Rossini Suite: 1. Marche—Pas de Soldats from *Guillaume Tell*, Act 3, no. 16; 2. Canzonetta—La Promessa from *Soirées Musicales;* 3. Tirolese—La Pastorella delle Alpi from *Soirées Musicales;* 4. Bolero—L'Invito from *Soirées Musicales;* 5. Tarantella "La Carita"—a sacred part-song for women's voices. Benjamin Britten incorporated the first phrase of the hymn "Come let us hear our cheerful songs," composed by Henry Hahee (1826–1912).

The ballet received a semblance of unity from the fact that all nine dancers remained on stage the entire time; those who were not dancing sat or stood at the sides to watch the various dances.[27] The three men were in white tights and short coats, not unlike Eton jackets, of different colors (black, brown, and blue) with high waistcoats of a contrasting design. The three women who danced the bolero had predominantly red and black dresses, the tarantella dancer had a saucy straw hat and a light bright dress, the dancer of the canzonetta was in white-spotted muslin, and the dancer of the tirolese had an elaborate regional costume.

For the opening march ($\frac{4}{4}$ time) there are three chairs on one side of the stage and four chairs on the other. Three women are grouped in the center while three couples dance around them. Presently, each couple is joined by one woman from the center; then the original woman partner retires to the center, leaving her cavalier to dance with the second woman. In this variation, Tudor emphasized interesting patterns as in ballroom dancing. He also used billowing lifts and a rich sense of stage space. He liked diagonal lines and legs flying in hitch-kicks. At the end the couples bow to each other, walk apart, and retire to the chairs. Another couple (Gerd Larsen and Hugh Laing) begins to dance a slow canzonetta. They move exactly alike as they execute light, flowing jumps, with *ballonnés, jetés, galops,* and *pas de basques;* the backs arch flexibly as their feet cut and beat. There follows a lively and complex tirolese, performed by Maude Lloyd and Antony Tudor. Large jumps with *tours jetés* and *grands jetés* as well as *ballonnés* and *tours en l'airs* characterize this difficult piece. The couple traverse the stage together and their movements swiftly mirror each other. A trio of women appears next for the charming Spanish bolero. They shape their arms in a classical Spanish manner, and toss their head in a fetching way. Tudor has the trio walking, almost hopping, on their pointes and taking one foot in and out of firm fifth position; the arms shadow the feet with full swirling motions. (Could this be how Fanny Elssler looked doing the "Cachucha?") Finally, a quick and challenging tarantella for Monica Boam and Guy Massey tops off the series of variations. This couple seems to sweep through its jaunty movement with the speed of lightning. Before the final tableau, the three men do *fouettés* upstage and two couples soar in big *grands jetés* straight downstage facing the audience.

Tudor was praised for his success in capturing the nature of regional dances in a very personal way. "The gay 'Tirolese' brimful of spirit, the 'Bolero' fascinatingly Spanish feeling, 'Tarentella' lightly Italian. Antony Tudor has gained this atmosphere by suggestion and not by a forcing of authentic national steps. Tudor, taking as a rule small parts invariably invests them with dignity and great perfection of detail."[28] More recently, John Percival wondered that this slim trifle of a ballet was able to survive until today, while other Tudor works did not: "Tudor's invention is so crystal sharp, the sense of style so acute that it still looks bright and fresh today."[29] Although the ballet may still look bright and fresh today, the dancers are not up to the power of the "old-timers." Maude Lloyd wittily described a particular partnering jump: "They can't do it any more. Everyone in the dance would titter. I would do a very rapid assemblé turning in the air and Tudor would catch me on the upbeat. He had to put one hand right under my bottom and push me up into the air. He held me suspended for a moment. I don't think they do it musically like that any more."[30]

When London Ballet and Ballet Rambert combined in June of 1940, after Tudor had left to pursue a career in America, *Soirée Musicale* was performed by the joint company. The first performance at the Arts Theatre was on June 21, 1940. It remained in the joint company repertoire until the combined company ceased to exist in September 1941. *Soirée* returned to Ballet Rambert repertoire at the Lyric Theatre, Hammersmith, on September 29, 1944. It remained in active repertoire until 1953, and again from 1962 to 1966. The first performance of the revival was July 25, 1962, at Sadler's Wells Theatre in London. *Soirée Musicale* has been restaged by many large as well as smaller regional ballet companies, including the Ballet Guild in 1942, the London Theatre Ballet in 1951, the Norwegian Ballet in 1957 and 1961, the Joffrey Ballet in 1959, the Scottish Theatre Ballet in 1973, the Juilliard School in 1982, the New Zealand Royal Ballet in 1985, and many more in the United States.

Tudor established his own company, the London Ballet, with Europe on the brink of war. He continued to play some of his most amusing ballets during this fateful period—*Judgment of Paris, Gallant Assembly,* and the new *Gala Performance,* which premiered with the opening of the London Ballet. Tudor said, "I knew there were people in the East End of London making a weekly pilgrimage to ballet in the West End and I was determined that they should have their own company."[31]

Looking for a suitable theater for his newly formed London Ballet took Tudor to many different sites. Finally, one year and six months after his departure from Rambert, the company opened in Toynbee Hall in White-chapel in the rather grim surroundings of the East End, close to the former stomping grounds of Tudor the clerk in Smithfield Market.

> It didn't take Tudor long to imagine how he could put it [Toynbee Hall] to use. With several important theatrical figures already on the Toynbee board, Tyrone Guthrie for one, the possibility of staging ballets there seemed good. . . . He rounded up his dancers and, adding Hugh's legacy from the Skinners to his small savings, was able to put the brand-new London Ballet into operation.[32]

True to his belief that a good ballet could be enjoyed by someone who knew nothing about the art, Tudor established a strong relationship with Toynbee Hall, a settlement house that had recently added a theater and large rehearsal rooms. Settlement houses provided poor neighborhoods with cultural amenities associated with the rich. Toynbee Hall, already fifty-three

Drawing: Toynbee Hall, 1938.

years old in 1938, was formerly an industrial school that had been bought by the church to inaugurate a workingman's college, where 2,000 students a year might take evening classes and "where undergraduates from Oxford and Cambridge meet and live among and join in debate with labourers and dockers."[33] The newly built Toynbee Theatre had 450 seats, a gymnasium, a canteen, a children's play space, and practice and rehearsal rooms. Apparently the utopian charitable purposes of Toynbee attracted American investors as well. The Carnegie Foundation donated 10,000 pounds to its enlargement in 1938.

Tudor could not have landed upon a more felicitous space, although the critic Jasper Howlett noted that most of the ballet audiences came from the West End and the suburbs and had long, tedious journeys to Toynbee.[34] An article illustrated with a picture of Tudor and a student announced, "The King and Queen will visit Toynbee Hall in London's East End. They will inspect a new block of buildings on which 50,000 pounds have been spent."[35]

Tudor took many of the Rambert dancers with him, all of whose relations to Rambert were now severely strained. Only Maude Lloyd was permitted to dance with the Rambert company, either because "Mim" could not survive without her, or because Maude's personality and bearing were so open and amicable. Tudor also brought with him his old dancing teacher, Margaret Craske, whom he had been working with for many years. She was known for her consistency and clarity as a teacher, a real coach for beginning as well as professional dancers. Miss Craske became his "technical advisor."

Since the ballets in the repertoire rarely demanded more than twelve in the cast, Tudor started out with a chamber company split approximately equally between men and women. In spite of the aristocratic nature of his own name, Tudor had an egalitarian bent: he gave Maude Lloyd and others leading roles, but there were no star dancers *per se.* In addition, all the dancers took the same salary, which was around five shillings a performance.

With great optimism, Tudor was ready to accept total responsibility for the choreography and production values of his own company; like Rambert, he would have the last word. The first season of the London Ballet opened on December 5, 1938, followed by fortnightly performances on Monday evenings and some Saturday matinees.[36] Billed for the first program were *Gala Performance, Jardin aux Lilas, Gallant Assembly,* and *Judgment of Paris.* A. V. Coton described this momentous event:

> At 8 p.m. on December 5, 1938, every disengaged dancer in London, most of the musical and theatrical press, the gossip columnists and what appeared to be about a thousand ardent ballet-goers filled the immaculate plush chairs or stood jammed in a tight phalanx at the back of the stalls. Contrary to the naïve expectations of certain parts of the audience there were few tall coats and backless gowns—and even fewer silk chokers and hobnailed boots. Ballet had come to the East End, not as an act of patronage to the East Enders, but for good practical reasons. A good theatre existed in the heart of the East End, a large part of the ballet-going public lived in or near the district, and the West End commercial theatres were out of reach on account of their excessive operational costs. . . . Disconsolate followers of the dancers who had got hopelessly lost on the way, kept on arriving almost up to the final curtain.

There was no room anywhere in the theatre and they waited in a huddled knot peering through a small window in one of the wing doors. . . .[37]

Lionel Bradley reported that a large audience attended this historical opening: Ninette de Valois, Mr. Beaumont, Margot Fonteyn, Robert Helpmann, Frederick Ashton, etc., and a great many of the audience who frequented the Wells and the Ballet Club. "There was some delay at the start, with hammerings behind the curtain. They were told afterwards that they had not been able to have a dress rehearsal and that they had barely got the scenery on the stage in time."[38]

None of Tudor's ballets presented during this season embodied old-fashioned conventions of classical ballet, such as a star ballerina with supporting cast, except *Gala Performance,* which was played as a parody. In London, only Kurt Jooss was driving the energy of ballet in directions similar to Tudor's. Janet Sinclair acknowledged that "they both had the compelling impulse to say something honest and truthful through stage dance without copying or stealing, without falsification for effect, without mental laziness and without thoughtless acceptance of convention." She went on to report that the first season was spread over twenty-one weeks and presented *Gala Performance* and *Soirée Musicale* (seven shows each), *Jardin aux Lilas* (six), *Gallant Assembly, Descent of Hebe, The Planets,* and *Dark Elegies* (four each). Two works failed to reach the boards: *Opera Ball* and *Pavane for Three Cavaliers.* Although presumably all the works were by Tudor, the Adagio duet from *The Gods Go A-Begging,* by Balanchine, was listed for the April 26, 1939, show.[39]

Apparently Tudor wanted to produce other new works for this spectacular occasion, but he ran out of money. He did re-create a simple duet for two women, "Joie de Vivre," which had been one of the *Seven Intimate Dances,* but it was not included on the London Ballet program.

Gala Performance (December 5, 1938)

Gala Performance was the only piece on London Ballet's opening night that had never been performed before.[40] Its savage satire of the stage manners of three ballet divas suited the taste of the sophisticated English audience, and Tudor's choice of Prokofiev's Classical Symphony perfectly synchronized with the light, but biting, mood. Tudor knew very well that his fledgling company did not have the technical prowess of its rival at Sadler's Wells. Therefore he produced a parody of ballet that would pretend extraordinary knowledge of the technique without its virtuosity. One commentator

Gala Performance, *1941. Nana Gollner and Antony Tudor.* (From the Performing
Arts Research Library)

quipped that "it was like watching an off night, a dreadful one, at the
Russian Ballet." According to Fernau Hall, "Tudor's conception of the temperamental foreign ballerinas is essentially an English one—it is a subtle
version of the Edwardian music-hall jokes about foreigners."[41] Several years
earlier, Rayner Heppenstall had asked in his column, "Why should such a
talent as Tudor be popularized? He may need a bigger stage. But he could
never compete with Massine or Ashton in playing to the Gallery."[42] It is
clear, however, that with *Gala Performance,* Tudor realized his potential to
create an audience-pleaser without trying to make a typical classical ballet.

Tudor reminisced about the initial inspiration: "I didn't base the Italian
on any Italian ballerina because I don't think I had ever seen one. But Italy,
to me, meant St. Peter's Basilica and the great sculptures. And so, she was
very slow, a great goddess, while the French was frou-frou, that was what
we associated with France. And with the Russian, I associated her with the
eagle, a bird of prey. And so rather than individuals, they were based upon
characters that I associated with those three countries."[43] Hugh Stevenson's
setting possibly outshone all the other productions he ever made. His rococo

costumes and the regal opulence of the setting backed by the pink and white fountains made it the most lavish decor he had yet achieved.

The first scene opens on several members of a ballet company with their backs to the audience. They are backstage, nervously warming up and preparing for their performance, doing *échappés,* bending *cambré* backwards so that their faces are to the audience. Gradually several more dancers enter the stage. Some tableaux have the mood and intimacy of Degas's paintings.[44] The corps de ballet doing their *pliés* nervously await the awesome arrival of the three guest prima ballerinas.

The conductor and dancing master prepare for the curtain to open. All the company turns to watch the Russian ballerina. She is perched forward, very arched over, like some kind of predatory bird; she holds her hands behind her back. Her attention to the corps de ballet centers on their total obedience to her. She orders one of the corps dancers to remove a necklace, even though she herself seems rather burdened with jewelry. Then she insists that they run through their steps. When the French dancer twinkles onto the stage looking like a French *patisserie,* all in pink satin and a shawl, they barely touch each other to kiss. The French dancer, the true carrier of the flame of ballet history, seems incapable of a thoughtful movement, so fluffy and superficial does she appear. The goddess of the dance, the Italian ballerina, enters in a black tutu, elegant, with high feathers signifying the heroic and regal. She moves slowly, like a sleepwalker, as if the floor will not hold her hallowed framework. The male dancers, cavaliers, are ready to kiss her majestic hand. The dresser holds up a mirror to the Italian, and suddenly, in one quick moment, she inexplicably slaps her maid for not concentrating on her every little wish. The scene changes rapidly to a true performance at the Theatre Royal.

During the second scene, Tudor spoofs the conventions that both distinguish and plague classical ballet. The dancers are now facing the audience against a backdrop rich in color, with typical fantasy-like designs worthy of a grand ballet. Eight women of the corps anxiously wait for their cue. Trying to be seen, they frenetically execute inside turns while making faces and mugging in front of the audience. The Russian dancer seems lost on the stage; she walks downstage, constantly peering at the audience as if she might find something there. After spinning several times, she continues idiosyncratically to bend over, her hands hanging down her body behind her, her head practically on the ground. Leaving the stage requires enormous courage on her part; she leaps off, fortunately caught by one of the cavaliers.

Ever so slowly, the peacock Italian ballerina walks in with her fountain of feathers waving. She nods to the conductor. She is always heading in one direction or another waiting to be partnered. But no partner could possibly attend to her exalted dancing needs. At one point she pivots slowly on one foot, the other in arabesque, flicking her wrists. She walks out and returns. None of her actions makes any sense whatsoever.

Flittering and fluttering, the French ballerina enters quickly, jumping all over the stage. The can-can in her hides behind every gesture. Light as a filly, she throws kisses everywhere. The Russian ballerina returns and does *fouettés* as in the Black Swan *pas de deux*. In between these scamperings, the male dancers jump high straight into the air with large *changements;* one lands badly, hurting himself, and limps off.[45]

The stage becomes a free-for-all when all three divas appear together. The Italian continues her stuporous walk; the Russian ballerina's head is constantly askew, in any direction the audience will notice, while her hands seem to drip lace from every finger. Finally, the French goddess pushes her partner off so that she can bow alone; she finds herself triumphantly caught in front of the curtain bowing by herself. When the curtain reopens, the furious Russian dancer, hands on hips, jealously steals some flowers from the others. After many upstaging antics, the ragged and aged fire curtain is brought down for the final applause.

One of the major problems with *Gala Performance* lies in the sensitive area of performance style. How can the dancer calibrate the extent of his or her exaggerated, comic gestures without overdoing them? Maude Lloyd was particularly emphatic about the way this ballet should be performed:

> Tudor did not want the ballet to be done any more without careful supervision. The dancers today just want to get the steps right. *Gala* is not a farce. If it is played for laughs, it's all wrong. It's very important to be serious, to take it all seriously. The role of the Italian ballerina suited me. I was very dignified, very quiet, I was supposed to be the most elderly, the most senior. At that time, the Italians were known to do a lot of balances, and took everything very quietly and calmly and very grand. The Russian role suited Peggy Van Praagh who was very forceful and strong, while the French role was taken by Gerd Larsen; she was jumpy and coquettish, very French. I always did the pas de deux with Antony. It was extremely difficult and I had trouble doing it. The solo was also not easy with its balances and turns. Tudor knew how to parody but we played it very straight. In the backstage scene the others were very noisy as it were. When I came out with my maid, I was not going to waste any energy. When she held the mirror for me, she wasn't paying attention and I called her to it and slapped her. Otherwise, I gave very little appearance of emotion. At another time, when the others are falling over themselves to be in front, we're looking at the audience and bowing. I give them a little elbow on either side to let them know I'm there and push them a bit.[46]

Lionel Bradley seemed to have some concerns as he commented on the opening night of this spicy performance: "There was much interest in the dancing, some of which was very difficult, but the whole effect was a bit scrappy and though there was a slight tinge of burlesque, I felt that the rivalry of the stars might have been more clearly brought out." He particularly appreciated Maude Lloyd's role, which he felt was the most successful. Concluding his review, he summed up, "As it stands the ballet is not of very great importance, but it could perhaps be revised into something better and more finished." Indeed, one month later, his review expressed delight and excitement as Tudor made a number of important changes: "The whole scene has been unified by the constant efforts of each of them to usurp the limelight from the other two."[47] There were other reservations about the ballet, however. For example, Cyril Beaumont wrote that "the music by Prokofiev as one of the modernist composers, seems out of the picture."[48] In addition he felt the transition from the backstage scene to the performance was too abrupt.

Gala Performance has received many performances from companies around the world, including the Royal Swedish Ballet in 1949 and 1957, the National Ballet of Canada in 1953, the Bavarian State Opera in 1960, the Deutsche Oper in Berlin in 1963, and the Royal Danish Ballet in 1970. American Ballet Theatre restaged *Gala Performance* in 1988 with the meticulous help of Hugh Laing (just before he died) and Sallie Wilson.

During this period of growth and productivity, Tudor's unflagging courage was remarkable. He fought to discover a strong choreographic process by experimenting with different theatrical genres. His compelling ventures in early television work have only recently been unearthed. He deliberately sought to become part of the Sadler's Wells Ballet, where he danced and choreographed operas.[49] What he really wanted in his early years was to choreograph for Ninette de Valois, but perhaps his difficult and domineering personality turned de Valois away from him. Or perhaps he simply could not compete with the charming and extremely prolific Frederick Ashton, who had already become a favorite of de Valois's.

From his earliest experiments with novel choreographic ideas, Tudor exploited the humor in his scenarios. De Mille often remarked that Tudor's comic bent overtook his tragic nature. Unfortunately, when he arrived in America, his penchant for the light, Wildean spoofing and irony disappeared. He became overly serious.[50]

CHAPTER FIVE

To America: The Journey That
Lasts a Lifetime

A luncheon discussion among Richard Pleasant, Lucia Chase, and Agnes
de Mille during a hot New York summer in 1939 resulted in an invitation to Tudor to come to America. They wanted him to mount some of his
ballets for a new enterprise that would soon become one of the great American ballet companies, Ballet Theatre.[1] It was the stunning debut in 1938 of
Tudor's London Ballet Company that prompted de Mille (who had left
England and returned to the United States early in 1939, at least partly
because of the imminence of war in Europe) to arrange this association and
to work on the creation of an American company dedicated to both new and
old choreography. The fledgling company, Ballet Theatre, would be struc-
tured rather in the fashion of museum galleries so that there would be dance
"wings," such as Classical, Spanish, British, American, and Black areas, that
represented aspects of America's melting pot.

Although Frederick Ashton had been first choice for an English choreographer, the group was pleased when Tudor agreed to replace Ashton. In
1939, Ashton was working with the Ballet Russe de Monte Carlo on the
ballet *Le Diable s'amuse.* On hearing that war was declared in Europe, Ballet
Russe canceled their season in London and made a quick decision to take
one of the last ships to New York, where the company opened at the
Metropolitan Opera House on October 26, 1939. Apparently Ashton declined the invitation from Ballet Theatre and generously suggested that
Antony Tudor and Andrée Howard replace him. Howard was developing a
strong reputation for her choreography with Marie Rambert.[2]

The European pilgrimage to New York was by no means new. Several
notable Russian choreographers had immigrated to New York in the 1920's
and 1930s. Michel Fokine in 1923, Michael Mordkin in 1924, and George
Balanchine in 1933 had made their way to the United States looking for fresh

opportunities, and they had found them. Along with Tudor, these remarkable foreign gentlemen assisted in the first flowering of American ballet. It is lamentable that the only English woman choreographer, Andrée Howard, quickly returned to London. Balanchine allied himself with Lincoln Kirstein, while Michael Mordkin (and to a lesser degree, Fokine) found a patron in Lucia Chase. Her wealth had originally come from the Waterbury Connecticut Chase Brass Company.

The new Ballet Theatre was an outgrowth of the Michael Mordkin Ballet. Michael Mordkin had grown up in the Bolshoi Theatre and had danced with the Diaghilev Ballet and Anna Pavlova. He founded the Mordkin Ballet in 1926, for which he choreographed several ballets, including a complete production of *Swan Lake* in 1927. His company, featuring some very accomplished artists, such as Hilda Butsova, Felia Doubrovska, Pierre Vladimiroff, and Nicholas Zvereff, also gave guest performances in Europe, but then disbanded. From the students of his New York ballet school, Mordkin re-established the Mordkin Ballet in 1937, and this became the nucleus of the Ballet Theatre company. His pupils included Patricia Bowman, Paul Haakon, Leon Danielian, Lucia Chase, Kyra Nijinsky, Katharine Hepburn, and Judy Garland.

The telegram Tudor received in September 1939 changed his life forever. War had broken out in Europe on September first, and Tudor needed money to finance his London Ballet. But it was not just the money that lured him to America; it was the opportunity to have his work seen there. He agreed to sail to New York and re-create several of his ballets for the Mordkin *cum* Ballet Theatre Company; he planned to return to London and rejoin his company later. His contract with Ballet Theatre was for ten weeks. Did he have any premonition that returning to an embattled Britain would become impossible? That he was, in fact, abandoning his infant London Ballet forever? It seems not. Tudor had arranged in his absence that the London Ballet would be directed by Maude Lloyd and Peggy van Praagh.

His correspondence with Richard Pleasant, the founder of the new Ballet Theatre, sheds light on this historic moment. *The Dancing Times,* a monthly English dance magazine, served as an intermediary source that cabled replies back and forth between Tudor and Pleasant. Tudor preferred to use airmail so that he could put his position more clearly. He explained in an August 15, 1939, letter that he could arrive by the beginning of September, prepared to rehearse the principals for two weeks and the corps for two and a half weeks:

> Naturally you would reserve my return passage in advance as it's absolutely imperative that I am working in London again by Monday 9th of October. I

have new ballets already scheduled and being announced. If it were possible to fix this, I would revive for you *Jardin aux Lilas,* Chausson, and Stevenson designer, or *Descent of Hebe,* Bloch and Nadia Benois, or I would bring the designs with me. Original work you would expect to be your sole property.[3]

Pleasant cabled back:

> Rehearsals 8 weeks minimum. $150.00 per week for work as a choreographer to give us non-exclusive right to ballets already produced. Performing salary, $125.00 per week. Must bring music and designs for "Jardin aux Lilas," "Dark Elegy" [*sic*], and "Gallant Assembly." We guarantee to mount two of the above. Passenger Cabin class two ways, to arrive October 15th.

Pleasant responded to Tudor's letter on September 5, 1939:

> New company in the process of forming. War has intervened. Perhaps you will be entirely unavailable. One thing is sure that America will probably become the only field for great endeavor in our field for at least a year. And what is the government stand concerning a young man who might become a general, but who is already a choreographer? Will they allow him to continue to be a choreographer and if so, only in his own country, or in America as well?

Tudor replied to Pleasant on September 12, 1939, in a telegram:

> Make passage arrangements and details concerning self and asst. Hugh Laing if possible. Laing paying own passage.

Pleasant answered September 18, 1939:

> For performances in January, bring music, designs, etc. Min. rehearsal period 8 weeks, $150.00 per week. For Hugh Laing, min. reh. period 8 weeks, $50.00 per week. Casting at our discretion. Performing period, $75.00 per week.

Again Pleasant cabled:

> Tudor, Howard, Passages arranged Sept. 21, 1939, on the Washington. Call for them.

September 30, 1939. From Pleasant:

> Am advising American Consulate that Tudor, Laing and Andrée Howard have 3 labor permits and bonds in readiness; advise me your departure.

Tudor replied:

> Cable American Consulate confirming three labor permits. . . . Regards sailing Washington—Definite.

The United States Customs Report for Tudor listed the following: One oil painting, twenty-one watercolor paintings, four pen-and-ink drawings, two pencil drawings, one photo, total value: $190.00. Shipped from London 10/1/39 on the S.S. *Washington.*

Tudor, Laing and Howard arrived in New York aboard the S.S. *Washington* on October 12, 1939. On the same ship were Anton Dolin, also to dance and choreograph for Ballet Theatre, Alicia Markova, Arthur Rubinstein, Paul Robeson, Paul Petroff, and "baby ballerina" Irina Baronova and her husband German "Gerry" Sevastianov, later to become the manager of Ballet Theatre. The boat was crammed, as many people thought it would be the last boat from England to America until after the war. Upon arrival in New York, Tudor and Laing were sent to Ellis Island because of a discrepancy in their documents: their passports had been issued to William John Cook and Hugh Alleyne Skinner, while their labor-entry permits were made out to their stage names, Antony Tudor and Hugh Laing. Tudor spoke of this famous first evening in New York as being quite an adventure.

In October, dance writers began to announce the Ballet Theatre's auspicious plans. John Martin wrote: "A New York season is scheduled to begin January 4, 1940, with 8 choreographers to begin Ballet Theatre. Michel Fokine, Adolph Bolm, Mikhail Mordkin, Antony Tudor, Andrée Howard, Agnes de Mille, Eugene Loring and José Fernandez. It is the idea of the planners of the company to represent all periods and styles of ballet, more or less in the manner of a museum, adding new works all the time."[4]

Of the choreographers engaged in preparation for the repertory for the new American company, four were Russian (Mordkin, Fokine, Bolm, Nijinska), three English (Tudor, Howard, Dolin), and one Mexican (Fernandez). It was not until December tenth that the first American, Agnes de Mille, was able to begin staging her ballet.

For the American wing, the Mexican-born José Fernandez was recruited without any difficulties. Pleasant had discovered de Mille's prodigious choreographic talent at the Guild Theatre in New York on February 12, 1939, when she and her group of twelve dancers performed some of the works she had created in Hollywood and London. Many of them were short concert pieces that would have to be extended to full-length works. One of these dances, *Rodeo,* eventually became the classic Western choreographic drama that Ballet Russe de Monte Carlo premiered in 1942.

In the fall of 1939, de Mille was busy choreographing the dances for *Swingin' the Dream,* an all-black musical version of *Midsummer Night's Dream,* starring Maxine Sullivan. The show opened at the 3,500-seat Center Theatre in Rockefeller Center on November 29. Unhappily for those associated with the venture—but happily for Ballet Theatre—it closed in ten days.[5] Consequently a theater was freed for Ballet Theatre's opening, and de Mille could fulfill her promise to do a "negro ballet to the music of Darius Milhaud."

De Mille was accustomed to working out her choreography on a small, carefully chosen group of dancers (in this case, two of them were Sybil Shearer and Marguerite de Anguerro) and then transferring the work to a larger corps. Some of the African American dancers she worked with in *Swingin' the Dream* joined the corps of Ballet Theatre. "In the Milhaud ballet, now called *Black Ritual* (*Obeah*) which premiered January 22, 1940, with sixteen Negro girls in the cast, Marguerite de Anguerro blackened her face and mingled with the dancers to direct dance traffic on stage."[6]

Into the cultivated, hybrid atmosphere at Ballet Theatre's opening nights during January 1940, Pleasant introduced Tudor's *Jardin aux Lilas, Judgment of Paris,* and *Dark Elegies.* At the Center Theatre all three ballets received their American premieres in less than two weeks, and all three were recognized by the public as works of exceptional artistic merit. *Jardin aux Lilas* and *Judgment of Paris* proved to be successes, if not total hits, while *Dark Elegies* could not quite achieve the high marks Tudor had hoped for; even the English critics had been dismayed by its theme and choreographic style. Early reviews of the 1948 opening season of Ballet theatre were optimistic and cheering, however. John Martin commented, "Mr. Tudor belongs to the extremely small fellowship of ballet choreographers who can take hold of the academic tradition and bend it to their uses."[7] Martin took some time before he came to grips with *Dark Elegies'* power. Eventually Martin acclaimed this work: "Tudor is one of the truly original and creative figures in the contemporary dance arts. . . . Only a choreographer of deep intuition and imagination would ever have conceived of making a ballet of Mahler's 'Kindertotenlieder.'"[8] Tudor liked to change his ballets, especially if the space's size differed from earlier performances'. In 1941, Tudor put the singer in the pit, and the review in the *New York Sun* praised the baritone in *Dark Elegies* for his lyric power: "He sang from the orchestra pit, better than last year when he was on stage, even more original and affecting than last year."[9]

Whatever the reviews of the moment, the dance and music worlds welcomed Tudor to American shores. For example, Peter Lindamood made several perceptive comments: "Antony Tudor not having been exposed to the special horror of Post War [I] Germany, has his canvas free of Teutonic frustration. . . . Let Mr. Tudor remain in this country. Here is the one man who can teach our young American choreographers the importance of content and complete clarity when staging a regional document. Right away he should attack the 'golden period' of New England in the 1840s, and shortly before. American primitives are jabbering stiffly over many walls pleading for animation."[10] Tudor's first important ballet in America, *Pillar of Fire*, although set in a later period, fulfills this wish to some degree.

An article in *Dance Magazine* by Anatole Chujoy optimistically praised Tudor's arrival in the United States: "The advent of Antony Tudor revives our interest and hopes and reassures our intuitive belief that our generation after all will not remain as barren as it had seemed these few years. Tudor is a genuinely talented choreographer of the first magnitude."[11] In the *Chicago Music News*, Neal Weilbel wrote, "Ballet American Style in a New Idiom: A Guide to Ballet Theatre." He paid tribute to *Jardin aux Lilas*, adding that "the roaring success of the season was *Judgment of Paris* the key to which is not found in Bullfinch's *Mythology*, but in the Paris of Toulouse-Lautrec. It's quite an unscrupulous comedy."[12]

An insightful comment from Richard Pleasant underscores the issue of wages in the arts at the time and especially in ballet. Richard Pleasant said that the minimum wage for the Ballet Russe de Monte Carlo dancers was $22.50 per week, whereas the minimum American Guild of Musical Artists (AGMA) wage for American dancers in Ballet Theatre was $45.00. "American Ballet is being threatened by the competition of the non-union Ballet Russe de Monte Carlo."[13] One can infer that Tudor, too, benefitted from Ballet Theatre's generous wage scale. He was making more money than he ever had in London.

In a copy of Tudor's 1941–1942 contract, it is written that he was supposed to give company classes if requested and would receive $75.00 per week instead of $20.00, and had the right to give classes in various schools. As *régisseur*, Tudor rehearsed the ballets and gave company classes. His dancing roles continued to be those that demanded better acting experience as well as a flair for character dancing, or regional dance forms and styles. In the program for the Chicago Opera, November 14, 1940, the following names are listed: Anton Dolin—First Classical Dancer and Régisseur of the Classical Wing; Eugene Loring—Demi-Character Dancer and Régisseur of the American Wing; and Antony Tudor—Character Dancer and Régisseur of the New English Wing.

For most of his American dancing career, Tudor tended to play character roles rather than *danseur noble* parts. Since he was a late starter, his ballet technique never achieved the level of technical prowess of his colleagues'. Besides, Tudor was not interested in becoming a great dancer. He preferred to be an actor-dancer and to imbue his roles with depth and interest. In addition to the characters he danced in his own ballets during the early years at Ballet Theatre, he is listed as playing the Dummy in Eugene Loring's *The*

The Great American Goof, *1940. Ballet by Eugene Loring with Antony Tudor in the top hat and Eugene Loring peeking underneath.* (Photo by Gjon Mili)

Great American Goof, King Bobiche in Fokine's *Bluebeard,* a Cadet in *Voices of Spring* by Michael Mordkin, and Pierrot in Fokine's *Carnaval.*

By June 1940, Tudor and Laing knew they would remain in America for a while.[14] Tudor must have encouraged Maude Lloyd and Peggy van Praagh to throw in their hats with the Rambert company, as trying to make the London Ballet work by themselves would certainly have been too much for them. Van Praagh wrote about storing the Tudor scenery and costumes in her parents' Hampstead basement: Tudor had left saying "I'll be back by Christmas."[15] During the war, ballets were performed in London during lunchtime, as the threat of bombings was diminished at noon. Rambert recalled, "Air-raid wardens and night workers of all sorts used to come and have a snack lunch and then an hour of ballet, after which, properly relaxed from the strain of their restless night, they could go to sleep for a few hours. It became so popular that we also did tea-time performances and then as many as four performances a day."[16]

Although Tudor created no new works for the initial American season at Ballet Theatre, he would eventually put together two inconsequential ballets, *Goya Pastoral* and *Time Table. Goya Pastoral,* created for a summer park performance, was not considered—and actually was not—a major work. He was invited to use the successful décor and costumes that Nicholas de Molas had designed for January's unsuccessful *Goyescas.*

Goya Pastoral (August 1, 1940)

Goya Pastoral tells the fable-like story of an aging marquesa who is infatuated with a young man. She supplies him with all her money until she loses at cards. This disaster, paired with seeing the Young Man, apparently dead in a suicide, hanging on a convenient tree, is too much for her nerves and she is carried off in a swoon. Whereupon the young man slips his head from the noose and lavishes the Marquesa's gold on a pretty peasant girl who sells grapes.

According to Pitts Sanborn, "Excellent in general were the dancing and the grouping, and Lucia Chase as the Marquesa and Hugh Laing as the Young Man were particularly well cast. As much can hardly be said of Tilly Losch as the peasant girl. As for the score, the orchestra proved more interesting than the somewhat insipid musical material taken from Granados."[17]

John Martin described the work as "a light hearted little ballet." He found the action slow in getting started, but once underway, Tudor's choreography, he felt, was "engaging and expertly done. . . . The choreography

Goya Pastoral, *1940. From left to right: Eugene Loring, Lucia Chase, Hugh Laing, and Tilly Losch.* (Photo by Werner Wolf)

makes no pretense at being authentic Spanish dance. Everything by Mr. Tudor is distinguished by his being able to give textures to his compositions. Tilly Losch, guest artist, was softly feminine and appealing, with little else."[18] Eight thousand people at the Lewisohn Stadium in New York watched several of Tudor's ballets, including *Jardin aux Lilas*. (One wonders how *Jardin* could have played to such a huge audience, given its modest birth at the tiny Mercury Theatre.)

In the *New York Herald Tribune*, Walter Terry remarked: "It turned out to be a visually pleasing work thanks to the scenery and costumes of de Molas and the neat choreographic designs of Mr. Tudor." Terry mentioned that the original ballet *Goyescas* had the engaging noisiness of castanets; but this production had a clearer plot and pleasanter movement. Terry declared that Tilly Losch appeared in dances that "hardly make use of her talents for sensuous and mysterious motions."[19] On the same program were excerpts from *Giselle, Peter and the Wolf,* and *Italian Suite* by Anton Dolin.

On their 1940 tour, Ballet Theatre brought sixty dancers, twenty-five ballets, and ten choreographers to Chicago. The vitriolic critic Claudia Cassidy commented on Ballet Theatre and particularly noticed Tudor's talent: "An essentially American ballet company, it has a cosmopolitan charm as well as its native exuberance. Its repertory is wide in range and polished in execution, capable of a magnificent revival of *Giselle*, and a virile *Billy the Kid*, of a cynical jest like *Judgment of Paris*, and of such deeply emotional ballets as *Jardin aux Lilas* and *Dark Elegies.* "[20]

In several clippings and chronologies, Tudor's *Gala Performance* is mentioned as first being performed in America on February 11, 1941. It became a successful and steady repertory piece. In a Chicago newspaper, the ballet is given a good review for its first showing on November 13, 1940, several months before its New York premiere.

The bustling activity of Ballet Theatre on tour and in New York seduced both Tudor and Laing into continuing their stay in the States. In an interview published in *The American Dancer,* Helen Dzhermolinska wrote:

> Antony Tudor is unshamedly frank in admitting that he came to America in quest of money. However he leaves unsaid what one suspects is the truth: that he felt the necessity of working at his craft somewhere away from the distressing nearness of war and that this weighed much more heavily with him than the desire for money.

Tudor exclaimed that he was bowled over by the technical abilities of American dancers but "finds their outlook upon art limited to a painful degree. Americans do not cultivate their parts with the feeling of the Continental or with the painstakingness of the English. Their interests are regretfully scattered among non-consequential affairs." In dancers, he admired, "first: quality of movement; second: a long slender neck; third: beauty. Accept his dictums you must."[21]

While Tudor was working feverishly on his first major ballet in America, *Pillar of Fire,* he was invited by Lincoln Kirstein to choreograph for the American Ballet Caravan in February 1941. American Ballet Caravan was an amalgam of Ballet Caravan and Balanchine's American Ballet Company. Kirstein originally founded Ballet Caravan in 1936 to function as a platform for young American choreographers.

Time Table (May 29, 1941)

Kirstein knew of Tudor's dramatic abilities and felt that a story ballet would enhance the formalistic Balanchine repertoire. Tudor was already quite pressed for time but agreed to accomplish a "little" work.[22] Although he had begun rehearsals for *Pillar of Fire,* he proceeded to prepare *Time Table,* an unprepossessing work about soldiers in a train station saying goodbye to their girlfriends. It may have given Jerome Robbins a hint for his sailors on leave in *Fancy Free* (1944). *Time Table* was originally performed in the Off-Broadway setting of the Little Theatre of Hunter Playhouse and accompanied by the NYA Orchestra under the direction of Robert Hufstader. The program (there was only one performance of *Time Table* at that time) stated that Lincoln Kirstein "invites you to two rehearsals by The American Ballet Caravan on the occasion of its departure for a tour of the South American Republics." They evidently wanted to give their repertoire a dry run in New York.

Zachary Solov, who at the early age of sixteen years played the gray-haired stationmaster, remembered that the ballet was not a huge success on its South American tour, but he noted that it contained the marvelous and distinctive characteristics associated with Tudor's work, such as a "very stark atmosphere, gestural movements with few classical arms, and getting through images of joy and sorrow with your acting abilities."[23]

Seven years after the premiere of *Time Table,* Tudor restaged it for the opening of New York City Ballet's independent ballet season. Previously New York City Ballet had appeared in City Center as a subsidiary of the opera season, but on January 13, 1948, "the house was completely filled and the auguries seemed good." John Martin explained that in order to celebrate the occasion, the program contained the first North American presentation (Hunter College was a one-night stand) of Tudor's *Time Table,* set to Aaron Copland's "Music for the Theatre," with scenery and costumes by James Stewart Morcom. Martin acknowledged the fact that *Time Table* was not a great work, but he stressed Tudor's talent for structure and atmosphere:

> It is merely the casual goings-on at a little country railroad station at the end of the first World War, when a train is due to arrive . . . and depart. Its only thread of a story is the farewell of a Marine sergeant and his sweetheart, beautifully played by Francisco Moncion and Marie-Jeanne. About them there move a pair of Marine privates, a couple of high-school kids, three young girls of a flirtatious turn (among them the always delightful Beatrice Tompkins) and a woman waiting for the return of her soldier lover. It is this return that lends a final poignance to the parting of the other lovers. The excellent Copland music has been around for quite a while and has withstood any number of choreographies. It has been beautifully used by Mr. Tudor, since he is a fine musician, even though it is not altogether his kind of music.[24]

Gisella Caccialanza, who played the role of the Girl, recalled working with Tudor and noted that the ballet contained many of Tudor's charac-

On tour with American Ballet Caravan, prior to sailing for South America. From left to right: Lincoln Kirstein, Gisella Caccialanza, Antony Tudor, Marie-Jeanne, and an unidentified man. (From the Performing Arts Research Library)

teristically difficult "non-dance" steps: "They're wonderful, when you think about it—just what you don't expect. But almost impossible. He used to give walk, walk, walk, then rise onto toe with no plié, no support. In his ballets you had to keep looking for opportunities to sneak up onto pointe."[25]

Like Balanchine, Tudor's broad range of talents enabled him to create dances that bridged musical theater and ballet. The movies of the 'thirties and 'forties acclimated audiences to a certain kind of show dancing that often hid behind the guise of elegances and smoothness as well as a highly demanding and strong ballet technique. Walter Terry wrote about the restaging of *Time Table:*

> It is slight material and Mr. Tudor has treated it in a manner which suggests a musical comedy interlude. There are moments eloquent in their communication of tenderness and desire, there are large-scale movements and impulsive gestures which define mood and character in masterful fashion. But there is also a good deal of superficial romantic dance and pretty obvious horseplay which, though pleasant enough, are hardly worthy of the distinguished choreographer's known skills.

Terry went on to say that the work was still important because New York City Ballet had no "American or idiomatic folk wing" at all.[26]

When Japan attacked Pearl Harbor on December 7, 1941, and destroyed the American fleet, all male dancers were subject to the draft, and fearful that their short professional lives might be cut even shorter. Both Tudor and Laing had been reporting to the British consulate in New York in order to see whether they would be called up by the British army. They never were.

Toward the beginning of the 1942 season, it became obvious that the treatment of Tudor at Ballet Theatre was not what it should have been. Margaret Lloyd disclosed that "Ballet Theatre is becoming Russianized. . . . For some reason the choreographer of this gem of ballets (*Lilac Garden*) and other important items in the repertoire was played down. Tudor received little mention in the advance publicity. In addition to being an ace choreographer, he took with distinction several of the mimed roles—King Bobiche of *Bluebeard,* the stalwart partner of the Italian ballerina in *Gala Performance* and the fiancé of *Lilac Garden* among them."[27]

Tudor was somewhat out of favor at Ballet Theatre. Perhaps they were disappointed by his slower pace in creating ballets. Even his old friend Agnes de Mille was critical: "He never under any circumstances started composing until the last possible moment. Once under way he could work longer and at a higher pitch than most and has for instance been known to compose a whole suite of dances in one afternoon. But at first he is slow, torturingly so."[28] Nevertheless, Tudor commanded considerable respect from anyone who studied with him. His dancers knew of his unfailing dramatic sense, sensitive musicality, and perfectly timed gestures, though they barely tolerated his disorganized methods of work.

Whatever the reasons for Ballet Theatre's treatment of Tudor, its management forces were then in transition. On March 9, 1942, Richard Pleasant announced his resignation. Economic forces in America were highly attuned to the difficulties and travails facing American troops in the war. Naturally, the arts suffered. Sol Hurok, the famous impresario of Russian dancers, took over the role of Ballet Theatre's company director. He believed that nothing could succeed like the panache of the Ballet Russe dancers. Hiring German Sevastianov to be his business and promotion manager, with Lucia Chase, he brought in Léonide Massine to work for Ballet Theatre during the season from 1942 to 1943. Massine received great attention both from the management and the public.

Displacements and troubles in America were moderate in comparison to disturbances across the ocean in England. The following letter indicates that after nearly two and a half years, Tudor could not or would not return to London to promote the work of his London Ballet. In the March 1942 issue of the English *Dancing Times,* Peggy van Praagh informed the dance public:

> The amalgamation of the London Ballet with the Ballet Rambert, which took place in June 1940, was dissolved in November, 1941. The activities of the London Ballet founded by Antony Tudor in 1938 have been suspended for the present. The Repertory, however, is still intact and it is hoped to resume performances at some future date when happier times prevail and conditions permit.[29]

Meanwhile, American dance audiences as well as dance critics recognized Tudor's abilities and creative potential. He was singled out by John Martin when he announced that the year's dance awards would go to Antony Tudor and the unique modern dancer Sybil Shearer.

> Fokine was the Romantic, Tudor is the neo-Romantic restoring the ballet from just as sterile an accumulation of falsity and to a human truthfulness that is far more potent largely because the psychology of the day makes a greater potency possible of attainment and acceptable in the dance art. . . . If Fokine had not existed Tudor could scarcely be working as he is working.[30]

With a strong sense of his own self-worth, like most people in show business, Tudor knew he had to fight for his rightful place at Ballet Theatre. But some of the demands he made at contract time lead one to surmise that either he harbored unjust feelings of being abused by the administration of Ballet Theatre or perhaps that his difficult youth prepared him to haggle unnecessarily in order to ensure that his talents were respected. In a letter from Tudor to Charles Payne on October 27, 1942, among other demands, he stipulated that the souvenir program must have a full-page picture of him, that he must be given the first male dressing room after Anton Dolin until the new *régisseur* was appointed, and that his name must appear as teacher of the company directly after *régisseur.*

No doubt the attitudes at Ballet Theatre were discouraging to Tudor. He felt that he had a great deal to offer them and that they lacked enthusiasm for his abilities. Ideas for new ballets did not come as quickly as for other choreographers, and, of course, Tudor needed a great deal of rehearsal time, which he felt he never had enough of in all his years at Ballet Theatre. From the beginning, Tudor traveled with the company on tour, rehearsing his ballets, learning new roles, and teaching company classes, responsibilities that certainly made it more difficult for him to choreograph. He also offered lectures on ballet and dance history as educational tools for organizations like the Friends of Ballet Theatre. Tudor's talent as a teacher and his former experience as a lecturer at Toynbee Hall in London were immediately apparent and frequently called upon.

The criticism at Ballet Theatre was silenced when in April 1942 a new Tudor ballet was presented. *Pillar of Fire* was arguably the masterwork of his career.

Pillar of Fire (April 8, 1942)

Tudor met his American muse, the remarkable young and passionate dancer Nora Kaye, when he first auditioned the company for *Jardin aux Lilas* and *Dark Elegies*. Eventually, he shaped her dancing sensibility the way Pygmalion created his beauteous goddess. When they first met, Tudor said that he could not take his eyes away from this raw young talent who had been dancing since age seven; she did five pirouettes on pointe. But that was not what touched him. He was moved by the emotive quality of her dancing, the significance and care she would offer each movement. She had a dramatic power and an understanding of her characters that came very naturally and could not be transferred easily to other ballerinas.

"Tudor first cast Nora Kaye in the corps of *Jardin* where, he said later, she impressed him with 'the way she embedded herself in the spirit of the thing,' and he singled her out to dance the haughty Russian ballerina in the first American performance of his satiric ballet, *Gala Performance.* "[31] Theirs became a powerful and important relationship as they worked together. Indeed, she played most of Tudor's major ballerina roles until Tudor left Ballet Theatre. When asked what effect Tudor had upon her life, Kaye responded, "From the very first moment, I fell in love with his choreography, and we became very close. He taught me everything. . . . Hugh Laing, who played The Man Opposite, Tudor and I began working on *Pillar of Fire* at Jacob's Pillow during the summer 1941, and the following season we were able to show parts of our work."[32] Kaye continued: "When Tudor

Pillar of Fire, *1942. Diana Adams.* (Photo by Alfred Valente)

started rehearsing with the three sisters and even before he started the actual steps—the choreography—he talked about our characters and the small town in which we lived. He even described the wallpaper of our house. It was so clear in our minds that we couldn't have done a wrong movement if we had tried."[33]

Pillar of Fire, *1942. Hugh Laing and Nora Kaye.* (Photo by Carl Van Vechten)

Before *Pillar of Fire,* Tudor's reputation as a choreographer in New York depended on his famous English ballets. *Goya Pastoral* and *Time Table* were slight offerings. The idea for *Pillar of Fire* had originally developed in London, when he slowly began rehearsing it with Peggy van Praagh and others. It was even announced in the *Dancing Times* in 1939. A letter from Tudor to Maina Gielgud of the Australian Ballet testified to the fact that, while he was still in England, Tudor had begun thinking about this ballet:

> The idea of making this ballet was back in England and would have had Peggy van Praagh as Hagar. And so the house might have been in a nice area of near suburbia in Nottingham? Buxton? Oldest sister is a Pillar of the local church, truly good and self-respected lady, still keeps the house in the most present- able and well-preserved way. The House across the way is seen through Hagar's eyes as mysterious, dark, sinister, and anyone going in or out of that house is invested with some sinister aura for we never know but that the Innocents and the other group are also only a part of Hagar's imaginings. Lighting goes from the warmth of a late summer afternoon through evening into night and the moonlight into the clear dawn with its intimations of spring and a new life.[34]

When Marilyn Hunt asked Tudor about the significance of the title *Pillar of Fire,* Tudor replied that

> God guided the Israelites out of the wilderness. And so Hagar was guided out of the wilderness. At least we suppose she was. That was the hardest title I've ever had to come across. I thought about it for months. Then we were sitting in the back of an open car on the way to Jacob's Pillow with a friend. And suddenly I jumped and said, "I've got it!" And they said what are you talking about? And I said, "I've got the title . . . Pillar of Fire." It was as sudden as that. It just came out of the blue. She was wandering around in no-man's land, wasn't she an outcast? And something brought her into civilization. And guess what it was. A Man? "You've got it."[35]

As Tudor explained, *Pillar of Fire* takes its title from a passage of Exodus dealing with the flight of the Jews from Egypt: "And the Lord went before them by day in a pillar of a cloud to lead them the way, and by night in a pillar of fire to give them light." The story of Hagar, however, is from Genesis 16: as handmaiden to Abraham's wife and at her command, Hagar lies with Abraham, and has his child; years later, she and her son Ishmael are cast into the wilderness, after Sarah and Abraham have a son of their own.

The contract that Sevastianov offered Tudor called the new ballet "I Dedicate." Much later, Tudor said "That wasn't even the working title. It was just something I put in the contract because there had to be a name. I decided on *Pillar of Fire* because Hagar, like Hagar in the Bible, was a lost soul. She thought she had lost a life of sexuality—and sexuality helps a lot of people out of their problems."[36] Tudor assembled dancers who he knew would understand his choreographic approach. They were inevitably to be known as "the Tudor group"—among them Nora Kaye, Maria Karnilova, Annabelle Lyon, and of course Hugh Laing. As Kaye mentioned, they started working out ideas for *Pillar* at Jacob's Pillow, a summer dance school and theater in Lee, Massachusetts, near Tanglewood. But when rehearsals began in earnest, voices were raised against it. Sevastianov and Hurok started calling it "Pills of Fire." After a runthrough showing of the piece, however, they decided to proceed.[37]

Laing felt Tudor "was wrong at the beginning about *Pillar* . . . he played the music for me in London and I told him he was a damn fool to try and use it."[38] In this case, Laing's usually impeccable advice turned out to be wrong. Arnold Schoenberg's music for the ballet, *Verklärte Nacht,* or "Transfigured Night," was composed when he was twenty-five, during a three-week September holiday. He had already set two verses by the German poet Richard Dehmel; he then took a melodramatic poem "Weib und die Welt" (Woman and the World), copying extracts onto the score as the basis for a heavily charged, chromatic work for two violins, violas, and cellos. Arnold Rose (Gustav Mahler's brother-in-law), leader of the Vienna Philharmonic, directed the first performance in 1903, and the audience was shocked. The daring narrative as well as the music gave the audience pause. Schoenberg had written program music for a string sextet! (Apparently, the players responded to one hissing audience by bowing, smiling, and repeating the performance as though encored.)

The poem, which was printed at the head of the score, tells of a man and a woman walking through a wood at night. She confesses she is pregnant, but her child will not be his, and she is tormented by guilt, as it is he whom she really loves. He comforts her, telling her to cast away her fears and that because of his love for her the child will become his. She feels redeemed by his love and forgiveness; as they walk on, the night becomes transfigured. The poem falls into five sections: Introduction, the Woman's Confession, the Man's Forgiveness, Love Duet, and Apotheosis, which suggest a ready-made form for a symphonic poem in one extended movement. The score is an uncanny depiction of moods in the retelling of the sad woman's tale: Her unmistakable guilt, her heart beating as she seeks courage to speak in that unhelpful woodland setting; the resolute strength of the man as he comforts her, and the eventual warm suffusion of their love.

Tudor worked with his own script in quite a different, more immediate manner. Although past experiences had haunted the heroine, he focused on the present and brought the intensity of her situation into vivid relief. Program notes indicate the following plot: "Hagar, whose elder sister is a spinster, foresees the same fate for herself. When the man she unrequitedly loves seems to show preference for her younger sister, Hagar in distraction gives herself to one she does not love. The resulting crisis, however, unites her with the man she loves." Thus, our understanding of the ballet requires that the younger sister assume a role of particular significance. In some ways, Hagar suffered a conflict not unlike Caroline in *Jardin*. At odds with social pressures and with her own best interests, she is driven by a powerful emotional attraction, in her case to an inappropriate partner. But the situation resolves itself when her love for the Friend overcomes both her shame and her guilt.

The ballet opens on a beautiful and illusory late nineteenth- or early twentieth-century scene created by Jo Mielziner, consisting of the façade of two houses at opposite ends of the stage, lit by the soft light of late afternoon.[39] The set, which indicates a home-dwelling downstage left and a house of ill-repute upstage right, projects a Southern aura. Others have noted that Appalachia might be a more accurate location. Tudor never wanted to be specific.

Hagar sits on her front doorstep, her body tense, her expression distracted, and her eyes uplifted. From the very first moment, when she lifts her hand and stares into the distance, the audience senses her deep sexual conflict. As Edwin Denby noted, each of Tudor's protagonists has peculiar movement signatures, "fragmentary pantomime suggestions" that imply inner tensions and that stylistically define his dancers.[40] Many such Tudoresque gestural comments follow in *Pillar*. When she raises her hand toward the side of her face, it seems to initiate the action of the ballet. People begin to enter as in a theater piece, introducing who they are and what they will be by their sparsely chosen gestures. The old sister, a spinster in long skirts, walks and moves purposefully, tautly, with necessity. The younger sister, in a childish short dress, bounces, spins, flicks her wrists, and tilts her head at whomever she tries to attract. Hagar, like most middle sisters, is caught between the temperamental shades of her elder and younger sisters. The length of her dress is in-between as well.

We meet the two men in Hagar's life-dream: the Man Opposite, who looks like a bartender (vest and rolled shirtsleeves); he slinks here and there, moving with jazzy syncopated steps, played originally by Hugh Laing in a stereotypically sexy way; and the Friend, played by Tudor, who will be her savior, dressed in a suit, correct in his bearing. He enters Hagar's house. Hagar always seems to be left out, like Thomas Mann's Tonio Kruger, looking in from the outside at the excitement of a party she is not invited to. In this case, she remains outside her own house, as well as the house opposite, in which people seem to be having a rollicking time. What does Hagar want?

The Man from the House Opposite appears, and there is no question any more about what Hagar desires. He lubriciously suggests what she needs by moving his body sensually. His leg swings open intentionally while his hands touch his thighs. Hagar does not yield the first time. A chorus of Lovers in Innocence—the women dressed like young women of the period, the men dressed in tights and a jacket—briefly dance in unison. Some critics have written that these corps moments do not work successfully, as the fact that everyone is doing the same movements sets them apart from the rest of the cast and one wonders why. On the other hand, one might view these dancers in a ritual manner, dancing as couples coupling in a charming, but intentional way. There is no question that Hagar is really the one puzzling and interesting person on stage, and she is there, in front of the audience, for practically the entire ballet.

It is difficult to follow the next phrases of music when Hagar's younger sister dances with the Friend. She seems to both encourage him to dance with Hagar and at the same time to twist him into a fickle web of games. Hagar dances in this trio, but always seems distant and out of control. Others act upon her. Her sister pushes her and she falls, as she does fairly often in the ballet; it seems to be a metaphorical gasp for what she is feeling and what is happening to her.

The first important duet then takes place between the Man Opposite and Hagar. She gives herself to him, desperately, and spends the rest of the ballet feeling guilty about her need for this man. Denby pointed to a particular lift, when

> she leaps at him through the air in a grand jeté. He catches her in mid-leap in a split, and she hangs against his chest as if her leap continued forever, her legs completely rigid, her body completely still. . . . The attention is focused on the parts of the body, their relation to one another, the physical effort involved in the leap and the lift, almost as if by a motion picture close up. And the moment so distinctly presented registers all the more, because it registers as a climax in the story, as a pantomime of a psychological shock.[41]

After her "Fall," she is greeted by her older sister, who responds to Hagar's misbehavior by looking at the heavens, her clasping hands held high above her head and then flung up and down. They are the only series of movements that misfire, as true as they seem, in the whole ballet. Gradually others enter the stage; Hagar's fright and shame seem to grow as the night darkens, and a quiet pause interrupts these whirling events.

The second scene of the ballet, although there is no set change, resolves what has preceded. As we re-enter this strange and intense environment, we glimpse an awkward Hagar as she tries to imitate the Lovers in Innocence. It is difficult for her to slip into a scene and move comfortably with the others. She is pathologically discontent "in her skin." When she re-encounters the Man from the House Opposite, she is maladroit in her attempt to attract him, her seducer. He pays her little attention, and when she drops down he drags her on the floor; she is truly a fallen woman. Her second and brief duet with him leaves the viewer exhausted and Hagar beaten in this masochistic exercise.

The entrance of the chorus, both Lovers in Innocence and Lovers in Experience, as well as the sisters, provides the scenic and metaphoric amplitude that will take Hagar and her new lover beyond the present. In the final romantic duet, in which the Friend seems to be constantly catching Hagar and keeping her from falling, she gives in to him and permits him to support her. In order to survive and to quell her tormented thoughts, she must acquiesce and find a way to join, to connect with the society she grew up in.

The concluding scene in *Pillar* is similar to the ending of *Dark Elegies,* moving from wild apocalypse to quiet resolve, even perhaps resignation. Here the couple walk quietly off the stage holding each other's hands, looking, not at each other, but into the future.

In *Choreography Observed,* Jack Anderson confirmed the point that "on the surface, *Pillar of Fire* seems a straightforward presentation of events arising from Hagar's fear that the man she loves does not love her. . . . But when one examines the ballet in detail, one finds that many points are ambiguous and it becomes surprisingly difficult to determine what literally happens."[42]

John Martin also observed that what takes place could be happening in the woman's mind. He asserted that the ballet moves with the pace and presence of a musical work rather than a narrative ballet: "By the use of cumulative formal processes, contrasting and conflicting minor themes developed in the shape of incidents, a tremendous emotional pattern is assembled and resolved ultimately into a welcome order that completely transfigures the original material." Tudor does the same with *Jardin.* He is evoking "inner states of consciousness" that are not told in a linear fashion. Martin affirmed that Tudor worked the way a musician does, "by interweaving different themes, and by using counterpoint." Often we are carried back in time. Gradually the piece emerges as the forms build themselves out of the material and the "spectator is carried along in the process." In the first part of *Pillar,* Martin continued, what happens on stage seems to be viewed through the eyes of Hagar. She is tortured, but she is also neurotic and responds to all the events as if they were terrifying. Occurrences become much larger than life as she listens to her own inner states of consciousness. "Women tuck up their back hair, tug a bit at tight collars, bow to each other in the most ordinary matter-of-factness, yet not one of these apparently idle gestures is without its emotional effect."[43] Here is where Tudor excells.

One may ask about the importance of Hagar's social class as well as her inner feelings, because her actions, as a bourgeois woman of the late nineteenth or early twentieth century, have enormous implications for her family and her place in society, as appearances must be kept up. Tudor himself, in 1982, explained to Anna Kisselgoff that "I never do a ballet that doesn't concern the bourgeoisie."[44] Tudor recognized Hagar, her older sister, and the boyfriend as familiar characters in his own experience: "I had to deal with the people I knew. . . . I didn't know princes and princesses." But these people have English references, not American. One of the problems with putting on this ballet today is that the American dancers who perform Tudor's ballets move with the assurance of athletes and the sensibility of youngsters. The women have not grown up wearing lengthy skirts, and the men have not grown up with "stiff upper lips." Consequently, the dancers have a difficult time re-creating important details of Tudor's "English" characters, not to mention his complex classical footwork and the expansive use of the torso in *Pillar of Fire*.

Marcia Siegel envisaged Tudor's message as a comment on social and cultural changes:

> Tudor details a community that is changing—in its structure and its behavior. The three sisters are survivors of a family in transition, living on in a once-genteel neighborhood that is becoming infiltrated by immigrants who haven't learned to refine their tastes or repress their emotions. . . . Time and again Tudor gave his character the refuge of tradition or asceticism—in *Lilac Garden*, in *Shadowplay*, in *Dark Elegies*, they draw back from experience and risk, into the safety of tribal codes.[45]

Nora Kaye recalled that, before its opening, only Tudor's cast knew how remarkable the ballet was. Antal Dorati conducted, and the show was a splendid hit, with thirty curtain calls. After the show, in a moment of uncharacteristic exuberance and braggadocio, Tudor yelled out to de Mille, "Agnes, I *am* Ballet Theatre."[46]

The reviews confirmed Tudor's greatest hopes. John Martin said: "Consider Antony Tudor. After *Pillar of Fire*, it is impossible not to consider him, and seriously, for here is no glib talent but something very like a new force in the evolution of dance." Martin goes on to praise Nora Kaye's powerful portrayal. "Against all these surface minutiae is the long sustained line of inner tension in the character of Hagar, so magnificently realized by

Nora Kaye that everything that takes place on the stage about her has value and heightened significance."[47]

In an interview, Sallie Wilson, who followed Nora Kaye as one of the most remarkable Hagars, reminisced about doing *Pillar* in Russia in 1966. "The Russians heard that the ballet was 'sexy' and insisted that we do an afternoon audition for the performance in case it was too suggestive. They were especially pleased that the costume had a high neck, and the movements did not seem to bother them. The ballet was able to be played and they found it very interesting."

Wilson also described one moment with Tudor in rehearsals. "He placed his finger in the center of my back, and told me to glow, to be alive in the chest, to maintain the stage." This swift, tactile observation helped her capture the essence of Hagar's movements. She added that his choreography for pirouettes was particularly difficult. "They mostly have no preparations. He would insist, just go up, wherever your heels are, don't rearrange your turnout. You press as if you were walking."[48]

Pillar of Fire remains one of the most important Tudor works of the American Ballet Theatre repertoire. Sometimes it is not performed with the integrity of style and approach that Tudor deserves. Other ballet companies have staged it, notably the Royal Swedish Ballet in 1962 and the Australian Ballet in 1969. In January 1990, it was offered in an impressive fiftieth-anniversary performance by American Ballet Theatre.

Just before Ballet Theatre's fall season at the Metropolitan Opera House in 1942, Tudor danced during the Ballet Theatre tour to Mexico. Massine created two new works with Spanish themes, *Dom Domingo de Don Blas* (1942) in which Tudor played the Viceroy, and *Aleko* (1942) in which he danced the role of the father of Zemphira (Alicia Markova). The reviews were not particularly favorable.

Almost exactly a year after the premiere of *Pillar of Fire,* Tudor presented another major ballet. Here Tudor delved into his love for Shakespeare and one of his most powerful romantic dramas. As usual, Tudor created something quite different from everything he had done yet.

The Tragedy of Romeo and Juliet (April 6, 1943), Incomplete; (April 10, 1943) Complete

One of the most shocking memories of *Romeo and Juliet* is the fact that Tudor could not finish this forty-five-minute epic by the date of the premiere. Hurok insisted that the ballet go on, even though the last two scenes were not choreographed. Sono Osato described this evening: "On opening night

Ballet Theatre in Mexico, 1942. From left to right: Nora Kaye, Jerome Robbins, Hugh Laing, Donald Saddler, and other members of Ballet Theatre. (Collection of Shirley Eckl Parker)

Tudor went before the curtain to explain to the audience what they were and were not seeing. The stir caused by his unexpected announcement died down quickly with the first notes of the music. . . . When the curtain fell, we held our breaths. At first there was total silence, then a burst of thunderous applause. Tudor stepped out again to thank the audience and offer his apologies for such an unusual opening."[49]

It is not difficult to understand the motives that induced Tudor to attempt to produce *Romeo and Juliet* as a ballet. To him, the vivid characterization and tense dramatic conflicts of Shakespeare's plays were highly attractive. Tudor had an early initiation into Shakespearean theater when he belonged to such dramatic societies as St. Pancras before he began to dance. He also achieved some success with his very first ballet, *Cross-Garter'd* (1931), based on the play *Twelfth Night*.

Interviewed by John Gruen about the birth of the ballet *Romeo and Juliet*, Tudor divulged:

> I heard a concert suite of the Prokofiev "Romeo and Juliet." I thought, I've got to use this music! Then I thought . . . it didn't have Shakespeare as I saw

Antony Tudor rehearsing Romeo and Juliet, *1943. Reflected in the mirror are Jerome Robbins and Alicia Markova.* (Photo by Alfred Eisenstadt)

Shakespeare. . . . I wanted it in the feel of the Shakespearean theatre—meaning, the feel of an open stage like the Globe. Then, I had to look for the music. I felt that the gentle folk melodies that are sprinkled through the music of Delius were terribly reminiscent of a lot of Italian melodies, and I thought they also had the colors of the Italian painters of the period. I could never find anything that really satisfied me for the last scene.[50]

Tudor also discussed the genesis of *Romeo and Juliet* with Alfred Frankenstein in the *San Francisco Chronicle:* "I had heard the music for Prokofiev's ballet and liked it very much but Fokine had been engaged to do it for Ballet Theatre. Fokine suddenly died in the summer of 1942. And then Ballet Theatre asked me to do it. I read an old fifteenth century Italian period piece that Shakespeare had used for his play. (Tudor refers to the writer Luigi da Porto, one of the play's most important sources.) Eventually I chose the music of Delius. After all, the music of an English landscapist should fit the mood of Shakespeare. I originally wanted sets and costumes by Salvador Dali because I had conceived the decor in terms of Fra Angelico and Dali handles those notions very well. However Dali's ideas did not harmonize with mine."[51] [Alicia Markova remembered that Dali "projected a giant set of false teeth supported by crutches for the balcony scene."][52] Tudor recalled less sarcastically that Dali provided a balcony in the shape of the prow of a ship and added, "However, the neo-romantic ruins of Eugene Berman were more in keeping with *Romeo and Juliet.'*[53]

Music critic S. L. M. Barlow called Tudor's choice of music "an opulent and Maeterlinckian score of Delius. There is a ton of plush in the music but so is there in the tale of the Capulets and the Montagues. Anachronism is Shakespeare's own oyster; and a score of contemporary Veronese madrigals would not accord with this essentially Dumas story which reached Shakespeare from Naples via the Palace of Pleasure."[54] Both Markova (*Markova Remembers*) and the critic Edwin Denby hailed Sir Thomas Beecham as the savior of the ballet. In the early performances, "the dancers were lucky if they heard most of their cues and they had a good deal of trouble getting their complicated sequences and shifts of balance to flow with the music. After a year's run the dancers still hadn't settled when Sir Thomas was invited in to conduct it. After eight performances with Mr. Beecham the dancers understood and performed the delicate aspects of the score that Tudor had counted on."[55]

Sono Osato shared a curious anecdote about the development of the score for *Romeo.* "Because of the war, it was impossible to obtain from England either full orchestral scores or the various instrumental parts for any of the pieces. After listening for hours to recordings, Antal Dorati sat down and wrote out the music for every instrument in the piece, transcribing each recorded note to paper. He also wrote and orchestrated musical bridges between the separate pieces, all in the style of Delius."[56] That was a labor of love!

Tudor demanded a good deal of his dancers. One day he came into rehearsal and asked the cast:

"What was the Renaissance all about? What were the key elements of that society? What was life like? And most importantly for "Romeo," how did they move?" There was an embarrassing silence. [Then Tudor proceeded to devise movement in keeping with the period. Osato remarked] We moved contrary to every movement of classical ballet, keeping our knees slightly bent, our pelvises tilted forward, and our necks curved to suggest the passive demeanor of women in Renaissance paintings. In contrast the men stood very erect, strutted more than they walked, and imbued every gesture with the fierce masculine pride of fifteenth-century courtiers.[57]

Eugene Berman's inspiration for the scenery and costumes was derived from Renaissance paintings, notably by Botticelli. All the critics sighed with wonder at the Berman sets, remarkably beautiful, architectural, and deeply three-dimensional. Structurally there were some important characteristics of the Elizabethan era that perhaps did not work. There were property drapes with two maids who sat and operated them; occasionally they did not open easily. There was also a question about the overdecorative funeral curtain. Yet the texture, the colors, and the wealth of rich and shiny fabrics projected a vitality and vivid luxury that made the tragedy of Romeo and Juliet all the more hapless.

The ballet opens with a prologue, in which Romeo and Rosaline engage each other in a brief scene. Though Rosaline is mentioned only briefly in the beginning of Shakespeare's play, Tudor chooses this moment to expose her qualities that serve as a catalyst and foil to Romeo's love for Juliet. The scene also balances the end of the ballet when the two lovers die together in the tomb scene. Osato, who created the role of Rosaline, recounted the way Tudor wanted Rosaline to suggest in her every move an abundance of sensual, teasing womanhood. And Romeo, played by Hugh Laing, was very interested in this display. Osato said:

> Romeo gestured to a black velvet curtain. In silence, I emerged in profile, with arms held stiffly to the side, a pose reminiscent of a Giotto painting. Demurely I rose on half-pointe, slowly rotating my thrust-forward pelvis, with a breath-like shift of my arms across my abdomen and breast to face Romeo. As he reached for me, I lowered my glance to the floor, rose again, and suggestively rotated my stomach away from him priggishly smoothing my hair with flattened hands. There was an elegant but sensuous rhythm to the movement based not on counts but on the elusive musical atmosphere. With minimal gestures my character revealed a propriety that thinly veiled a tantalizing wanton.[58]

The stage promptly fills with Montagues and Capulets jousting and battling. Tudor gives the large male corps numerous jumps and beats in curious contrapuntal rhythms, and keeps the stage alive and in motion with beating jumps to second position, often in canon. The women's dresses flow this way and that as they kick their legs from one side to the other in pointed *jetés* and *ballottés*. They wear headpieces (Osato had to make hers at the last moment from two *Scheherazade* bracelets) and swing their skirts, one arm weightily slung across the stomach; they push into the hips with gliding steps. The whole stage seems to billow like a ship. Occasionally Tudor designed pictures in which several individuals executed different movements at the same time; this activity creates a slightly syncopated atmosphere and at the same time lends definition to different groups of people. The arms and hands were never afterthoughts in Tudor's work, especially not here. He not only used them as part of every movement, but also worked them into the spatial design of the patterns. It was important to Tudor, who had studied character and period dancing, to carry across the mood of an era, in this case the Renaissance style, with its push into the hips and tilt back for the gliding women and its oppositional and erect stance for the men.

In the ballroom scene, the male dancers use Basque steps, with hands above the head, clicking fingers, and stomping feet.[59]

The ballroom episode is heightened by the girlish timidity of Juliet and her childish dependence on her nurse; although the Nurse in this version has less emotional and farcical value than in other versions of the ballet. For the characterization of the Nurse, Sono Osato described a particular rehearsal moment:

> Like Fokine, Tudor demonstrated so brilliantly that no dancer seemed to reproduce his movements with equal clarity. He worked with Lucia Chase one afternoon on a single instant of the Nurse's part, showing her how to evoke the sticky heat of the day by rising and gesturing with her arms as if to pluck her heavy skirts from her buttocks. He did not want her actually to move the costume away from her body, but to isolate the familiar act of doing so by combining the sensations of heat, clothing, and damp skin in her gesture. Lucia eventually captured the essence of the moment, but she never looked quite so hot or sticky as Tudor had.[60]

By bringing very few people into full focus, Tudor distills and compresses action. For example, the Montagues, Romeo's parents, play very slight roles. The meeting of Romeo and Juliet happens subtly and with discretion as Juliet dutifully dances with her fiancé, Paris. Romeo mingles with other women, as he has come to the ball "to examine other beauties." When Romeo and Juliet finally dance together, the meaning of their relationship unfolds with urgency. Overcome with emotion, Romeo falls to the floor and Juliet bends over him. Her movements are silkily coy but essentially serious. She turns away; he reaches for her as if to capture a kiss once more, "Give me my sin again." A piercing doom-like chord sounds; they both turn and fall to the ground as their duet ends. At this point, Tybalt discovers Romeo's presence. There is another fight, and Romeo flees. The music during all of these opening scenes, Delius' "Over the Hills and Far Away," has a richness and pageantry that delicately supports the presentation of Shakespeare's characters and plot. The violins are soft, at times lush, and do not overwhelm the action. Tudor keeps the dramatic action tense and expectant; the stage remains alive as couples move almost in slow motion, gesturing to one another.

The balcony scene, central and intimate to the argument of the play, represents the kind of restraint and emotional selection Tudor is known for. Romeo's mercurial, sometimes enigmatic character emerges most clearly in this scene as we watch Juliet with the "mask of night on her face" observing him. Romeo moves swiftly in and out of shadows and hiding, "with night's cloak to hide me from their eyes." Deborah Jowitt noted, "Antony Tudor, alone of the many choreographers who have tackled Romeo and Juliet, keeps his protagonists separated during the Balcony scene despite the urge to create a showy ecstatic pas de deux."[61] The miming of the lovers in the balcony scene forms a central static panel in a garden alive with life, swarming with the figures of Capulets and Montagues bearing torches and the young girls of Verona stealing through the arbors until the viewer actually breathes the fragrance of the Italian night. All about Romeo and Juliet there is movement while they stand rooted to the spot, conveying their rapture not through dance but through gesture.

For both the balcony and the betrothal scenes, Delius's "The Walk to Paradise Garden" is used. After the betrothal scene, another fight erupts in the street; the men do not wield swords and daggers. They leap at and lift one another; some fighters doing daring arabesques to the changed mood and pace of the "Eventyr" music. Tudor chooses a stylized and unusual movement vocabulary, accenting judo-like positions and shadowboxing. The death of Mercutio and Romeo's killing of Tybalt happen as if in a dream, as if being imagined after the fact by Juliet.

Romeo's farewell to Juliet in her bedroom does not occasion a grand *pas de deux*. Rather, Tudor emphasizes their quiet determination to remain together despite what would seem their imminent destruction. In the play, Juliet's haunting thoughts predict Romeo's fate: "Methinks I see thee now, thou art so low, As one dead in the bottom of a tomb." Here we see a Juliet dancing as if drawn with every movement to Romeo's being. And then her closeness centers her. Romeo falls to the floor; Juliet goes to him. When Sono Osato rehearsed the part of Juliet in the bedroom scene for Markova, Tudor created a poignant moment, seating Juliet on the edge of the bed, her feet on the floor, side by side:

> Her left arm crossed her breast with the other hand touching her right shoulder. The right arm then crossed her abdomen, and the other hand went to her hip. In one moment the body turned towards the audience into another two-dimensional Giotto-like position, with her legs in a wide turned-out second position. Her elbows bent into the torso, with hands upraised, the flattened palms turned outwards. The head tilted to the left, eyes downcast, a perfect image of an idealized Madonna. However, at the time of performance the

movements were changed. Apparently Markova crossed her feet in ballerina fashion and rounded her arms and hands.[62]

Markova's position, according to Osato, was lovely but became a nineteenth-century classical pose rather than the haunting quattrocento image Tudor had intended.

The music, "Brigg Fair," accompanies the preparation for the wedding of Juliet and Paris. In the early part of the ballet, the prevailing qualities of Juliet are a kind of playfulness coated with a gentle innocence. Here Juliet begins to understand; she becomes increasingly distraught and detached from her surroundings. Before her nurse and her parents, Juliet refuses to marry Paris, but her handmaidens hold before her a cloth of gold, and Juliet, resigned, slips her arms into it. Her father, Capulet, places a crown on her head. When the Friar enters, she falls into his arms; her savior who will administer the poison from which she will be reborn.

After the procession to the tomb, the scene in the vault of the Capulets shows Juliet lying on her bier, her hands pressed together on her breast. Believing Juliet to be dead, Romeo drinks a vial of poison and falls at the foot of the bier. Juliet revives from the drug, and rises. Here, unlike the play, Tudor brings the lovers together, as in Luigi da Porto's version. Romeo joins Juliet; they walk quietly downstage toward the audience. After one last passionate lift, he goes to the tomb and falls to the ground. Realizing what has happened, she returns to him, takes his dagger, pointing to herself ("This is thy sheath, let me die"). With her back to the audience, she rises onto her toes, plunges the knife into herself and slides down to the floor over her lover's body, touching his face before expiring.

In the epilogue, the palace scene from the prologue reappears, and the two families gather on either side of the archway in this moment of grief.

Tudor has created a reverie, a meditation on the play of *Romeo and Juliet*. His choice of music, perhaps unexpected, suggests a different approach to the theme. In Tudor's view, the heavy drive of the commonly heard Prokofiev music imposed a cumbersome narrative and punctuated quality on the ballet. Delius's music hints and makes remarks and gives a sense of enveloping the movement. It floats through the piece, reinforcing our interest in the ballet. Of course this is the way Tudor uses music. Occasionally we feel that the dances or the dramatic action foretell the music, such as when the couple falls to the floor, before a loud, prescient, doom-like chord. In some ways Tudor casts a new light on his dancing Romeo; he characterizes him as brash, not hot-blooded in his killing of Tybalt, but almost moralistic, and indignant at the death of Mercutio. He is more changeable than driven by brutish, animal passion. Hugh had developed his personality into an ardent, impetuous person from his role as the cruel young man in *Pillar*. Denby perceived that Laing's portrayal of Romeo in the ballet "is never quite frank; he is like an object of love, rather than a lover."[63]

Indeed, Tudor did not create a realistic and physical ballet where overwhelming anger leads to murder or where graphic scenes of lovemaking are depicted. Tudor preferred to distance us, by means of a poetic mood. Markova's Juliet is like Ophelia, or Giselle, if you will, fragile and wispy, delicate but deliberate, not at all childishly silly and bouncy. Juliet is drawn to Romeo, who secretly captures the stage and creates a magnetic field around him. As if to accentuate a certain mysteriousness in Romeo's character, Tudor moves him in and out of shadows, appearing and disappearing. One is almost relieved at the end of the ballet when Juliet can finally hold onto him, as though it were only in his death that she might be fastened to him. Like several of Shakespeare's early tragedies, the play hangs, as A. C. Bradley noted, on a thread of accident that does not permit the lovers to wake at the right time in the tomb. When Tudor gives his Romeo and Juliet a last chance to be together, we recall, by seeing them dance, their tragic vulnerability and remember all the more painfully Markova's frailty and Laing's foolhardy recklessness. Apparently Nora Kaye took over the role in October 1943, and John Martin commented that she dealt less in line and contour than Markova and more in emotional impulsion.

Simplified and condensed, the ballet unfolds with a certain quiet insistence. Tudor has treated the lovers with flexibility and fullness. The completeness and self-surrender of the love between Romeo and Juliet is clearly rendered, but there is hardly a moment when we are allowed to think that permanence or happiness is part of Nature's plan. The rapidity of the scenes (there are about ten of them) and the ephemerality of the dance form itself mirror the transience of their happiness. By keeping two ladies of the Capulet household on stage all the time, pulling the curtains, and introducing scenes, and by filling the stage with families and neighbors, Tudor reminds us that the lovers are part of a larger canvas and that their brief existence is enhanced and enriched by their Renaissance world.

Eight days after the complete version of *Romeo and Juliet* premiered, on April 10, 1943, Tudor was honored by Ballet Theatre with the presentation of a laurel wreath and a parchment scroll including the signatures of every one of the dancers and staff in the company. In a sense, Tudor was crowned king of Ballet Theatre not four years after his 1939 arrival in New York.

Contemporary critics seemed disappointed in Tudor's idiosyncratic treatment of the play. John Martin found that it was "a play without words rather than a ballet, and in many places the words can almost be supplied to the action as it unfolds."[64] Irving Kolodin, who was normally a music critic, agreed with Martin: "He has undertaken nothing less than a literal representation of the Shakespearean drama in forty minutes of dancing."[65] Edwin Denby criticized the movement: "As dancing, the piece is not as rich in variety or as exact in placement as the serious Tudor pieces." But he praised its "interesting attempt to keep the dance more continuously in motion," and after seeing it again, noticed that it stood up extremely well as a drama.[66]

In 1971, Marcia Siegel confirmed Denby's opinion when it was restaged for American Ballet Theatre. "Antony Tudor's *Romeo and Juliet* is a beautiful ballet. An extraordinary ballet. I think it's the best new ballet I've seen this year, even though it's really a revival. . . . I don't know when I've seen a ballet with such clarity and distinction of detail."[67]

Romeo and Juliet was restaged by the Royal Swedish Ballet in 1962 and by the American Ballet Theatre in 1971.

If the critical response to *Romeo and Juliet* was initially disappointing, Tudor's next ballet was greeted with more approbation. *Dim Lustre* appeared a swift six months after *Romeo and Juliet*.

Dim Lustre (October 20, 1943)

Tudor was comfortably settling into Ballet Theatre's hectic fall schedule—teaching ballet classes, dancing character roles in Fokine's *Bluebeard* and Eugene Loring's *The Great American Goof,* and choreographing new works. Much of this activity took place on tour. There is some material in the nature of a journal evidently designed to help Tudor remember the circumstances of *Dim Lustre*. (Unfortunately, the journal found in the New York Public Library Dance Collection is not dated.)[68]

Tudor called these three typewritten pages *"Dim Lustre,* Another Story." Apparently while Ballet Theatre was touring the north Pacific states, Tudor had lunch with a well-known ballet teacher, Mary Ann Wells. The young Gerald Arpino and Robert Joffrey were her stellar students in Seattle. She told Tudor that she had discovered a piece of music that was "made for Tudor," and Tudor remarked that "no woman could have been more right." When Tudor listened to Richard Strauss's "Burleske," he wrote, "I was enthralled more possibly than she could ever have hoped, and immediately, the next morning, I went to Lucia and told her that I had the music for a ballet I would like to do."

Once again Tudor had to cope with Lucia Chase's predilection for Russians. Tudor found out in the spring that Massine had been engaged to do a new ballet for Ballet Theatre at the Metropolitan Opera House. Chase

Dim Lustre, *1943. Hugh Laing and Nora Kaye.* (Photo by Fred Fehl)

responded to Tudor by saying, "And dearie you know there isn't any more time available." So Tudor had to put his inspiration in the back of his mind and let the "bright idea germinate." In those days Ballet Theatre did train tours, which gave everyone plenty of time to think. When he returned to New York, Tudor received a phone call from Chase, saying, "Antony, isn't it wonderful, now you can do yours." Apparently Massine had reneged on the new ballet, and Tudor had the go-ahead for his "bright idea."

In these three brief pages Tudor rewards us with some rare memories that illuminate the influence of his youth on *Dim Lustre.* "My mind jumped to the picture of some book shelves in the corner of the sitting room when I was a lad. These book shelves held bound volumes of *The Boy's Own Paper* and more recently had been joined by volumes of the *Strand Magazine.* " In addition, Tudor mentioned "memories of visiting the Café Royal at the lower end of Regent Street, and of a dance card my mother had kept as a souvenir from her courting days which came from some place I remember being 'Mr. Burry's Ballroom.' The geography of the piece then had to be developed onto the structure of the music, of its shapes and emotions and from this idea gradually evolved an ideally romantic couple at a ball during the course of which one of the other's memories were prompted into an alive influence through the chance happenings that could so easily happen in such a situation."

The lush and flamboyant Strauss music fired Tudor's thoughts and dictated how he would structure the scenes and his movement. Unlike with his other works, he whipped off the ballet in several weeks. Tudor confided to Jennie Schulman of *Dance Observer*. "'At one time I could never arrange waltzes. I had an absolute horror of them until *Dim Lustre.*' Since his major dancing in the Met production of *Die Fledermaus* consists of waltzes we can consider his phobia laid to rest."[69]

The choreography of *Dim Lustre* is rich and full in design, and, although the stage is constantly in motion, the ballet retains its formal clarity. The dancing catches the satirical spirit of the music, which whisks and bluffs like a series of light jokes. Tudor always enjoyed a good snicker at upper-class behavior, especially at the sexual games people played; *Gallant Assembly* and the later *Knight Errant* attest to the fact, however innocent of guile the protagonists believed they were.

The ballet opens on a lush red backdrop rich in Viennese ballroom motifs and chandeliers, against which five orange and red waltzing couples, all looking very much alike, float and whirl through their predetermined steps and patterns. Costumes for the women, although strapless, have a weighty ruffle that drops along the bodice and softens the torso line. The neck and chest are ringed with three layers of small pompoms. The sweeping skirts are also bordered by pompoms while the long line of the neck and head is continued with feathers poised on top of the crown. Elegant short jackets, bow ties, and trousers give the men music-hall, dancing-partner airs. The opulence of the costumes was used as advertising for evening-gown styles at Bonwit-Teller in the *New York Times* on November 21, 1943, with drawings of Nora Kaye and Hugh Laing: *"Dim Lustre,* a Swinburnean ballet, Fantasy of the Edwardian Era!"

Greetings and partings begin the ballet, entrances and exits that carry our eyes to different areas of the stage. Since the dancers move by so quickly, like a blurred photograph, one wonders if what one has just seen was really there. A Belle Epoque couple twirls in, and focus is on their intimacy and aplomb. One immediately notices Nora Kaye's face, the tilt of her neck and expressiveness in her upper body, but especially the dramatic value of her head movements on a very vertical torso as her skirts swirl.[70] Tudor insisted that his ballerina keep a stiff upper back, that she "remains a lady, no matter what." Hugh Laing plays the adoring, adulating cavalier. But is he? The stage vibrates with the florid strings and percussion of Strauss's music; all the couples waltz in a circle and do large jumps (*grands jetés*) straight downstage to the audience, as if to say, "watch us fly, we are important people living in a precious time." (Hands clap, the lights go out.)

The Gentleman with Her kisses her on the shoulder, triggering a transformation. The lights go down, the music stops; the lights return to reveal the lady facing her double and moving exactly like her. In fact the stage seems to be divided horizontally by a huge mirror. How easily and quickly her mind takes her back to her youth when another young man, It Was Spring, kissed her on the shoulder. These momentary flashbacks are interrupted by the ballroom scene when the other dancing couples interpose themselves both spatially and psychically, tugging these day-night dreamers back to the present.

As in all of Tudor's dances, we watch the play move on; the movements, like the diagonal and circular patterns of waltzing steps or the balletic quick *bourrées* from one side of the stage to the other, do not necessarily draw attention to themselves. Yet we do rivet our eyes on the language of his gestures. When the Lady with Him drops her handkerchief, Tudor cautioned Leslie Browne in the 1985 re-creation, "You don't know you're going to drop it," and cited Luise Rainer's performance in the film *Masquerade* as an example of what she should aim for. This time it is the man's memory that folds back to the moment when he fetches his first love's handkerchief, Who Was She? choosing her from among others who look like the Three Graces. Tudor mentioned that when the young man sniffs the handkerchief, sometimes the audience laughs. Is he blowing his nose? Is he going to sneeze? Is he smelling her perfume? Tudor did not think it mattered if there were titters; what concerned him was the dancer's focus on the gesture, that it be convincing and clear. The couple's memories have been jogged, and more characters from the past appear.

Tudor cannot resist presenting a few types worthy of late nineteenth-century drawing-room plays. Could these happenings be fantasies? The gentleman remembers an elegant older woman, She Wore Perfume, whom he met on the deck of a luxurious steam liner. She is a sexy and distracting charmer, but he is way too young despite his yearnings. The boat sways, and the wind is cold. In a parallel manner, the lady recalls He Wore a White Tie, a sophisticated gentleman (the perfect role for Tudor) who reminds her of a more passionate and riper love affair. He Wore a White Tie is a heartbreaker, dashing and confident. There is just a hint of brutality in his bearing. What follows, according to Hugh Laing, is that the young man comes out of his reverie. "He looks at Nora. 'Do you have that same perfume?' he thinks. 'Will it be too much?' And Nora is thinking of the older man who once swept her off her feet. Then we look at each other. Something has changed. Our memories have changed us."[71]

The dancing itself is one of the subjects of this ballet, not just an expression of the couple's soft waves of memory pushing back the moment. Apparently the stage crew waited in the wings of the old Metropolitan with cups of water for Kaye and Laing, who recalled that "'it was the most tiring ballet I think I've ever done. It was very quick, a step a note.' But Laing added that probably it was not the difficulty of the ballet that tired them, but rather the 'intensity of emotions.' People fall in and out of love because they are social-dancing, holding, and touching each other. And at the end of the ballet, they keep on dancing, just as in the beginning. Nothing on the face of it seems to have changed. But as Hugh noted, when the couple look at each other, everyone knows that they have become different people."[72]

Tudor strengthened the scenario and his character's relationships by using subtle spatial compositions. He architecturally placed the protagonists in very sensitive physical areas; sometimes by detaching the lead couple from the solid force of the waltzing group, at other times by carefully placing them in an isolated spot with their doubles or former lovers who, of course, have an importance and a connection unknown to the others. The use of a double to differentiate time in the hero or heroine's life was a device that Martha Graham and other modern dancers began to employ after *Dim Lustre*. In a Proustian or Bergsonian way, Tudor chose particular experiences to highlight—a whiff of perfume, a taste, a sudden sensual recognition that recalled and recaptured times past. Tudor understood the immediate power of sensual memory. But in the theater, memories can be difficult to evoke. Thus, Tudor invented another device that served his purposes: he periodically turned the lights on and off, as quickly as blinking an eye, bringing on or off the protagonists' youthful figures of memory.

The ballet generated a lively response. One critic joked that "in *Pillar of Fire* the lovers can't, in *Lilac Garden* the lovers are prevented, in *Dim Lustre* they won't." Edwin Denby offered some reservations: "The rhythm or dance style in both the dream and the reality have no difference choreographically and the dancing in the piece is not as rich in variety or as exact in placement as the serious Tudor pieces: what the ballet has instead is an interesting attempt to keep the dance more continuously in motion."[73] John Martin called *Dim Lustre* a psychological episode, "*Strange Interlude* in the Lilac Garden. Mr. Tudor has never before been so perky. He has taken the slightest possible theme and teased you into going also with him."[74]

In *Dim Lustre,* subtle and eloquent acting enhances the refined mating-game themes. Gervaise W. Butler declared that *Dim Lustre, Pillar of Fire,* and *Lilac Garden* are all about sex. She noted, "It takes a major imagination to draw such variations of the tragic, the frustrated, the comic from the same idea."[75] She added that Martha Graham and Antony Tudor had a lot in common.

Tudor remarked that he thought a good full-evening program of his ballets would place *Jardin aux Lilas* and *Dim Lustre* back to back. He noted that "they're both about the same length and the music is not so dissimilar." Also, "the people were the same"; their upper-class concerns, heavily guarded by etiquette and social graces, brought them together. He had told Maude Lloyd, his original Caroline in *Jardin aux Lilas,* that *Dim Lustre* was a ballet that she could very easily perform.[76] Both ballets briefly and delicately reveal the dismantling of lovers' memories and imaginations: *Jardin* is a romantic and painful view of those strangling moments of passion; while *Dim Lustre,* with a witty sophistication about youthful affairs, is a nostalgic satire.

In 1985, when Tudor and Laing first re-created *Dim Lustre* for the American Ballet Theatre tour to Los Angeles, "The dancers hadn't a clue to motivation and only thought of steps. It had been nearly thirty years since Ballet Theatre had performed it. Gradually, by the time the company got back to New York and had also prepared *Dark Elegies* and *Jardin aux Lilas,* they learned some of the important qualities of dancing in a Tudor ballet. Then the ballet succeeded."[77]

Dim Lustre remains in the repertoire of American Ballet Theatre.

Barely two weeks after the opening of *Dim Lustre,* on November 1, 1943, during an all-Tudor evening program, Tudor was photographed with a joyous expression receiving a gold watch from Sol Hurok for his contributions to the company.

Tudor's enthusiasm for America and for the future of American dancers was demonstrated in an article he wrote for *Musical Courier,* "America as the New Home for Ballet Tradition," in which he praised Ballet Russe de Monte Carlo and Ballet Theatre. He saw their work as an important influence in encouraging the appetite of Americans for ballet. Americans "had come to look upon ballet as an alien, rather odd, effete and very 'fancy' form of theatre art, possessing little with which the average American could feel sympathy. But a few people and espeically S. Hurok . . . succeeded in keeping the ballet companies going until something of American understanding of ways of thinking and feeling, found their ways into some of the ballets that constitute the repertories of the companies now regularly performing for American audiences."[78]

But enthusiasms were easily extinguished. As Linda Szmyd perceptively remarks, the political atmosphere of Ballet Theatre was thick and complex: "All choreographers were equal, with no one artist in a more powerful position than another. Rivalries were unavoidable. So were attempts to gain more power. Tudor's titles between 1940 and 1948 list him variously as choreographer, dancer, régisseur, ballet master and artistic administrator."[79]

Szmyd said that Oliver Smith, who took over the direction of Ballet Theatre with Lucia Chase after the "Hurok Era" in 1945, recalled that Tudor always "wanted more power." In 1946–1947 an artistic advisory committee was formed with Agnes de Mille, Tudor, Chase, Smith, Jerome Robbins, Henry Clifford, and Aaron Copland. "Most of the meetings degenerated into 'bone picking' between the artists, each of whom had their individual axes to grind."[80]

It was also during this period that George Balanchine worked with Ballet Theatre. His association began in 1943, when he restaged his *Apollo*. Later he created *Theme and Variations* and contributed to a production of *Giselle* with Tudor and Dolin. Tudor's relationship with Balanchine was "cordial," according to Oliver Smith.

After creating three major ballets in fairly quick succession, Tudor evidently had some difficulty in moving on. Never a facile worker, he also refused to repeat himself. His next ballet was born painfully two years later.

Undertow (April 10, 1945)

Could two ballets be more dissimilar in style and inspiration than *Dim Lustre* and *Undertow?* Although only two years apart, one resembles a soufflé; the other, bitter herbs. Just ahead lay the end of World War II, in Europe in May of 1945, and the detonation of the atomic bombs over Japan. The news of the war, the cost in human lives, and its destruction and terror had their impact on Tudor. While the earlier *Dark Elegies* was strongly influenced by the devastation caused by Fascist raids on Guernica in Spain, *Undertow* presented a metaphor for Tudor's own milieu. Modern warfare, the Holocaust, urban poverty, moral revulsion at the cost of lives in Europe and Asia set the tone and psychological backdrop for a unique ballet that dealt with the destruction of a young man's psyche.

According to Ballet Theatre's Charles Payne, the plot for *Undertow* was passed on to Tudor by John van Druten, "perhaps because he realized it could not be employed for a play on the legitimate stage. Where but in the dance theatre could one have then enacted such ghoulish sequences?"[81] Payne suggested that *Undertow* was one of the most English ballets of Tudor's early choreographies. Although the setting, designed by the Chi-

Undertow, *1946. Nana Gollner and Hugh Laing.* (Photo by Baron)

cago artist Raymond Breinin, may have been intended to depict an American city, it more closely resembled the people and the slums of Liverpool or Newcastle.

The program for the American Ballet Theatre season in 1968 reprinted an excerpt from the program of the Philharmonic Symphony Society of New York on October 5, 1946, the first concert performance of *Undertow:*

> The ballet concerns itself with the emotional development of a transgressor. The choreographic action depicts a series of related happenings, the psychological implications of which result in inevitable murder. The hero is seen at various stages, beginning with his babyhood when he is neglected by his mother. . . . The frustrations engendered by this episode are heightened during his boyhood by his sordid experiences in the lower reaches of a large city. He encounters prostitutes, street urchins, an innocent young girl, a bridal couple, dipsomaniacs and a visiting mission worker whose care and friendship he seeks. The emotions aroused in the abnormal youth by these episodes . . . revulsion, rage, terror, loneliness, fear of domination . . . result in climax after climax reaching a peak of murder of a lascivious woman. It is

only when he is apprehended for this crime that his soul is purged by the tremendous relief that is his at the realization that he will no longer be called upon to endure the anguish of being a misfit and an outcast among his fellow men.

In 1956, the program notes shrank considerably to include just the following:

The ballet concerns itself with the emotional development of a transgressor. The choreographic action depicts a series of related happenings, beginning with his babyhood when he is neglected by his mother. The frustrations engendered by this episode are heightened during his boyhood by his sordid experiences in the lower reaches of a large city, and only resolve themselves in his murder of a lascivious woman.

When Ballet Theatre made its tour to London in 1946, Tudor told the English critic Fernau Hall that *Undertow* would be a flop for the critics, as it had been in the States: "They can't take the sex." To some extent he was right about the critics, but the audience reception was quite astonishing:

Some members of the audience found it incomprehensible. . . . It was like magic, black magic; the cheers on the first night recalled the halcyon pre-war days when Massine was producing such ballets as *Symphonie Fantastique*. Strangely enough there were only three performances in London, but the word spread like lightning. "A whole new audience of writers and artists flocked to see Tudor's ballets, especially *Pillar* and *Undertow*.[82]

Tudor suggested that the mixed response might have derived from confusion about what was going on during the ballet: "It was exactly what I intended and what I expected. If audiences are to gain the fullest enjoyment, they must be trained, as well as the dancers, to do some of the work. To everyone in the audience, it meant something quite different, which I think is good. Like life itself, a work of art has many facets and can be understood in many ways." Tudor explained one of the last moments in *Undertow:* "For example a child's balloon floating away could mean a loss of innocence among psychologists, to the realist, local color, to the poet, man's aspirations."[83]

When asked about the origins of the ballet, Tudor recalled:

I got the idea for *Undertow* because I wanted to do a ballet about a murder. So I had to find a new angle, an appropriate approach. No, not an appropriate approach, but an intriguing approach, something the audience would look on as a novelty. There were three murders in *Romeo and Juliet,* but they were too Shakespearean to count. I would say that my ballets are practically anything else but sensual. People who know me very well would admit that I was not a sensualist. I agree with them. I wouldn't get an emotional feeling from seeing *Undertow*. I would react purely intellectually.[84]

Yet, for the public Tudor did create a sensual ballet, sometimes harsh and repulsive. "His treatment of sex nearly knocked the socks off the English audience; it was too close to the bone."[85] The graphic *pas de deux* at the end of the ballet reminds us of the *pas de deux* in *Pillar* when Hagar unashamedly got what she wanted. But it seemed that the shock for all audiences in watching *Undertow* was the Transgressor's murder of the prostitute during the sexual act. Tudor developed it as if it were an anatomy lesson, as if he had read about a homicide in the newspaper and then traced the murder back through its Freudian meanings. His psycholanalytic approach to murder attempted to establish the moments of unconscious memory that impelled the murderer to action, as if in a case study. Unresolved Oedipal conflicts, for example, would create repressed guilt and perhaps cause the person to react violently. Tudor chose names from classical mythology for all of the characters in *Undertow,* except the Transgressor, perhaps because they provided original insight and theatrical truths. They also offered a symbolic map or pattern, rather than a particular narrative experience.

When Tudor commissioned William Schuman to compose the music for *Undertow,* the only music score ever commissioned for any of his ballets, he did not explain that the dancers would have "symbolical names."[86] Most of the reciprocal suggestions and remarks were relayed through the mail, by telegram or by telephone. Schuman complimented Tudor for being sensitively precise in his description of the length of sections and the specific emotional climate he hoped the music would help him to produce. "The ballet is to take place in a city in Central Europe, an unidentified city. It is the kind of night in which one feels moisture on the palms of his hands and in which one turns a corner only to pause a minute and look back before proceeding."[87] For the first scene of the ballet, Tudor wrote: "It should be very calm, with occasional, sudden outbursts which subside right away. These outbursts continue spasmodically for a little while. Then you hear them coming closer together, and finally there's a serenity. That's where your first wonderful melody comes in."[88]

Schuman mentioned that he had no idea that the opening scene was to be a "choreographic presentation of a human birth." Nevertheless, "the character delineations given to me by Tudor, and the precise timing and atmosphere which he hoped the music would produce, were all so absolutely clear in emotional intent and so intellectually exact that there was no problem of communication between choreographer and composer."[89] And for the last, "getaway" section of *Undertow,* Tudor asked Schuman to compose a four-minute musical essay representing fear. Schuman found this task appropriate and perfectly possible to execute.

The action seems to take place in the mind of the Transgressor. The scenery and symbology suggest that Tudor wanted to remove his anti-hero from realistic situations. The characters in the ballet, except for the protagonist, have Greek or Roman names that point to contexts of a sexual nature.

In the prologue we see Cybele, the Earth Mother, in a kind of spot lighted Olympian solitude. Present also is Pollux, who was Cybele's mate. Mythologically he existed as the immortal half of the Gemini, in addition to being a great boxer. In order to share immortality with his brother, Castor, he spent one day on Olympus while Castor was in Hades, and one day in Hades while Castor was on Olympus. The goddess mother Cybele, in a red tunic, lies face up in the lap of a kneeling male. Her son is in a fetal position nearby on the floor. Her arms moving in an abstract gestural manner, she hides her face, which looks anguished as she rises. She lies down again while the boy nearby begins to roll on the floor close to his mother Cybele and father Pollux. Cradled in his father's arms, he hunches over in a contraction, his legs in *plié* like a paleolithic man. The mother walks around him. She holds him in her arms and nurses him but, significantly, moves away to Pollux; the boy gestures, reaching with his hands, high and outstretched to the side. Watching his parents, he backs away from them upstage into the shadows. Cybele remains on stage and dances a quiet solo, with *piqué arabesques, bourrées* and balances, before she departs.

The scene becomes a street in the slums of a Central European city. Statues of great winged horses dominate the nearby square. An exquisite if street-smart Volupia, the personification of sensual pleasure, places herself in the same spot where Cybele had danced. Volupia's movements resemble Cybele's, although they are lewder and more suggestive. She wears a scarf that is interwoven through her hair.

The Transgressor, now a young man, enters with Aganippe, a little girl tripping mincingly along on her toes and carrying a ball. She represents the virtuous nymph who inspired all who drink from her fountain. The Sileni, elderly and drunken companions of Dionysus, resemble satyrs in their inability to keep their hands off the nymphs. The Transgressor moves toward Volupia, who looks at him contemptuously; she prefers to catch one of the nose-picking Sileni for herself. The Satyrisci, the young worshippers of Dionysus, strut and preen with full movements and large jumps across the stage.

Aganippe returns with Nemesis, a young woman who regards all the action with numb fascination. Volupia lures an old man to her side; he fawns and cringes in the wake of her exaggeratedly seductive advances. The Transgressor looks on while sucking his thumb and watching the battering of the old man. His feelings, flustered and complex, drive him to offer himself to Volupia and he is immediately rejected. She slaps him to his senses. At one point the Transgressor circles the whole stage, his head dropped, and his feet dragging. He is truly an outsider.

An overcheerful busybody, dressed in a Salvation Army cape and hat, tries to collect an audience for a prayer meeting. (Sallie Wilson noted, "The Salvation Army lady was supposed to be rapturously good, good in the sense of being pure, but later, when she suddenly becomes sexually attracted to the Transgressor, and latches on to him in a lecherous manner, she shocks herself.") Of course he shies away from her.

Pudicitia and Ate, the hideous creatures who would lead all men into evil, appear as sorry and unlikely volunteers for prayer. Although Ate looks petite and innocent, one immediately focuses on her suggestive and salacious expressions. In direct contrast to Ate, an idyllic-looking wedding couple (Hera and Hymen) bounce across the stage with light, jumpy steps, she wearing her bridal veil and holding a bouquet, he looking excited, content and surfeited. The innocent demeanor of the bridal couple quickly evaporates when three unappetizing Bacchantes, like distant memories of the three goddesses in *Judgment of Paris,* in aprons and high-button shoes, shoot lascivious glances in all directions. They reel around one another and open their mouths in silent bawling.

A scene begins innocently as Ate flirtatiously taunts the Satyrisci. She is playfully pushed from one man to another, but she senses danger as they begin to throw her around. Unable to free herself, she is carried off by two men holding her arms and one carrying her feet. It is a chilling rape scene. These streetlife characters loom large for the anti-hero playing out his sad psychic voyage. When Ate re-enters alone, the Transgressor accosts her. One of the most moving episodes in the original production centered on Alicia Alonso's predatory, insalubrious characterization of Ate, especially this moment when she is almost choked by the Transgressor but manages to wrench herself from him.

Finally Medusa enters, the mythical monster who petrifies those who look upon her. She wears a belted tunic and is played by the same ballerina as Cybele, the Transgressor's mother. (In the original production, Diana Adams played Cybele and Nana Gollner took the role of Medusa.) The Transgressor grabs hold of her and partners her. Others, such as Pollux and his new woman Pudicitia, enter and look on. In this ballet, someone is almost always watching someone else. In the gaze lies either an accusation or lustful thoughts. One sees these kinds of stares in Racine's seventeenth-century tragedies. The Transgressor cringes when another girl comes toward him. Is he forbidden to love? Medusa kneels and the Transgressor places her hands as if in prayer. He assumes the same contracted, bent-over crouch he held in the Prologue. While they are moving together, he falls to the ground, and Medusa towers imperiously over him. Often they cross nervously back and forth on diagonals; she pirouettes, he stealthily catches her. Her sexuality is so overwhelming that she begins to overcome his inhibitions. She brings out feelings in him that he has only experienced vicariously. They both fall to the ground and repeat some of the same movements that Cybele and he performed in the Prologue birth scene. Now, he is alone with a woman for the first time. Medusa lies on her back on top of the Transgressor. Almost imperceptibly, their bodies molded together, he suddenly chokes her to death during the sexual act.

The scene changes to a backdrop of a misty city with the winged horses of the square rising into the sky. The Transgressor, barechested as in the first scene, is shown terrified and alone. He begins a long solo in which he realizes the depth of his horrific experience, where sex, evil, and death come together. He scampers here and there, looking as if he is trying to escape from his hideous past. Some of the city people stroll past indifferently. Ladies pass, carrying oranges that spill out onto the stage. He glances at Aganippe playing with a balloon; she stops and stares at him. He smilingly moves toward her. But she points fiercely at him and releases the balloon from her grip. The other characters join her and point at him. Like a condemned man, he walks off the stage slowly.

Almost every critic remarked that the ballet was shocking. Some found the movements and the direction of the ballet at odds with the psychology of the protagonist. What Tudor usually accomplished with such refinement, they felt, he neglected in *Undertow*. Edwin Denby offered his explanation: "Because *Undertow* lacks such a physical release of opposing forces, it remains intellectual in its effect, like a case history, and does not quite become a drama of physical movement. The trouble is, I think, that the decisive initial scene, presenting a bloody birth, brilliantly shocking though it is, does not seem to be a part of the hero's inner life, it is not placed anywhere in particular."[90] The impact of the ballet seemed to grow on the public, however.

One year later in London, when Nora Kaye replaced Nana Gollner as Medusa, Fernau Hall wrote: "There were large numbers in the audience who found it tremendously exciting. The role of the Transgressor danced by Hugh Laing was played with a vitality and psychological truth new to male dancers: he portrays the torments and warring impulses of an adolescent with disquieting precision."[91] Tudor's understanding of the pubescent male was as bitter as it was insightful: "This little boy should look like a Picasso

Blue Period boy—skinny and sensitive, bedraggled and feeling un-wanted."[92] As Keith Roberts, a current Ballet Theatre dancer who has played the role, described him, "He's constantly rejected. It's as if there's an adding machine in his head. He keeps adding things up, and as he does so, he carries this huge burden."[93] As the young man discovered what the world was like, this "sentimental education" acquainted him with a bizarre series of mostly unappetizing women, starting with rejection by his cold and indif-ferent mother, Cybele. Tudor paraded before the Transgressor a kind of rogues' gallery of grotesque female personalities. Even Aganippe shifts into a mean child. But the men are no better: "Everybody he meets turns out to be a fake," Alicia Alonso commented to Fernau Hall.

The falsity and hypocrisy of these characters provoked in the Trans-gressor first a deep unhappiness and then a psychosis. Although Tudor read numerous psychological texts in preparation for *Undertow*,[94] he also used devices common to literature. For example, he showed a series of images that suggested the whole personality and mind of the character, so that we know not only what that character is thinking and feeling but also his unconscious motives. Many of the characters are treated in a broad, satirical manner. All the women have appetites too large to accommodate, and the men seem to be victimized in one way or another. For Tudor, perhaps, male aggression and female lust are at the root of violence. The characters in *Undertow* are reminiscent of the personalities in expressionist dramas by Wedekind or Capek. As in their expressionist plays, Tudor's message paints a bleak, futile picture; all the incidents lead to one, usually violent, ending. One could ask if Tudor might have seen Kurt Weill's operas or Alben Berg's *Lulu,* whose scripts describe the evil and decadent impact of postwar city life (Berlin during the Weimar Republic) as well as the driving force of fate.

Another significant point made by Fernau Hall is that Tudor's choreog-raphy displayed his proclivity for Japanese *Noh* theater. For example, he constructed *Undertow* in three sections. The first two-thirds of the ballet is only partially danced by the Transgressor, who watches the others move; in the second section the whole theme builds up to the final dance, where the Transgressor relives the totality of his life. Originally, during the change of scenery to the second scene, many surreal props were thrown on the stage, but they were later eliminated. According to Hall, "The whole long final dance was tremendously complex. At the very end, Hugh is going to give himself up. He's relived his life, he's now prepared to die, he's understood why he committed the murder."[95]

Jack Anderson acknowledged that Tudor's social comments are ap-proached innovatively:

Undertow is crammed with incessant nervous movement suggesting urban restlessness. Everyone constantly rushes to and fro, creating the effect of a low fever which in the sex scenes mounts towards hysteria . . . an objective hysteria which can be noted in many cities. Danger may lurk anywhere. No wonder the Transgressor's habitual stance is a wary half-crouch.[96]

The dancing in *Undertow* contains many hints of modern dance technique and occurs in fragmented passages neatly sewn together. Tudor's movement vocabulary takes on a completely new range, with an emphasis on the theatrical power of gesture. As Edwin Denby observed:

Indeed one keeps watching the movement all through for the intellectual meaning its pantomime conveys more than for its physical impetus as dancing. Its impetus is often tenuous. But its pantomime invention is frequently Tudor's most brilliant to date. The birth scene, an elderly man's advances to a prostitute, the slaps of the prostitute, an hysterical wedding, drunken slum women, several provocative poses by the hero's victim, and quite particularly the suggested rape of a vicious little girl by four boys, these are all masterpieces of pantomime, and freer, more fluid, more plastic than Tudor's style has been.[97]

Consequently, the characters in *Undertow* play their street-smart types with more sweep and breadth as well as sincerity and commitment. Tudor's dramatic style affirms that the movement flows from an impulse directly connected to the action. The ballet is wrenching. When the single male Transgressor figure leaves the stage, the only real person in the ballet and the only one we identify with relinquishes himself to his destiny, which he has only just begun to understand.

At various times Tudor changed the ending of *Undertow,* always dissatisfied with the last impression that the ballet projected. For example, during one production Tudor asked the cast to come on stage and point at the Transgressor. In another version the Transgressor ends the ballet by lying down and curling up in a fetal position. For the 1992 production, Aganippe pointed at the Transgressor, and he slowly walked up the stage and off.

American Ballet Theatre restaged *Undertow* in 1967, with the Dutch dancer Steven-Jan Hoff in the title role of the Transgressor; and in 1979, with Johan Renvall. Renvall reassumed the role of the Transgressor in the successful American Ballet Theatre 1992 staging by Sallie Wilson. Deborah Jowitt applauded Renvall's searing performance: "Renvall brilliantly conveys this horrified sense of a gentle man who can't even fully reach out of his tense held-in posture, but whose soul records every cruel or erotic gesture until it can store no more images and bursts."[98]

CHAPTER SIX

Disenchantment in New York

*T*hough Tudor was fully employed at Ballet Theatre, like many of his contemporaries he often took on additional projects. He felt as comfortable choreographing for the musical theater or opera stage as he did for the ballet stage. And he continued his interest and involvement in educating dancers during the summer at Jacob's Pillow. In England, he had often produced work for opera, musical comedy or revue, and television. In America, Broadway offered better financial compensations for these diversions than one received in the concert world.

During this complicated time in his career, Tudor made the dances for the musical comedy *Hollywood Pinafore,* with lyrics by George S. Kaufman and music by Arthur Sullivan. It opened in New York on May 31, 1945, just six weeks after the momentous opening of *Undertow.* His Broadway debut received some raves, if brief. The show satirized Hollywood's cash consciousness, and Tudor's choice of music digressed from the *Pinafore* score, using instead *The Mikado* and the *Yeomen of the Guard.* In addition Tudor was commissioned to do the dances for another Broadway show, *The Day Before Spring,* which opened November 22, 1945.

In most of the reviews, the critics mentioned that Tudor's dances in *The Day Before Spring* did not enhance the show's thin plot. Wilella Waldorf spoke of the two dances by Tudor as not "top Tudor," in spite of Laing's performance.[1] Robert Coleman also criticized Tudor for not "giving Mary Ellen Moylan and Hugh Laing a chance to display their talents. The dances are not off Antony Tudor's top shelf."[2]

Beginning in 1946, Tudor headed the ballet department at Jacob's Pillow, where professional dancers and students alike took classes and performed. In an interview with Gregory Carmichael in *Tricolor Magazine* in New York, Tudor shared his views on Jacob's Pillow:

I should like to start a sort of summer festival for ballet in the manner of the popular music festivals. There is so much that could be done with a little help with the great deal of talent that seems to be going to waste now. New ballets could be tried out there with very little expense. There is a great future for ballet in America. It is just now getting on its feet and I want to have my share in seeing that it gets a good hold.[3]

He went to the Pillow almost every year and loved being there, rehearsing, taking time to ponder and to be "in the country." There he teased out a number of charming ballets, such as *The Dear Departed, Les Mains Gauches, Ronde du Printemps, Trio con Brio,* and *Little Improvisations,* suited to the students and dancers who spent summers at the Pillow.

Despite these artistically fulfilling side interests, the situation for Tudor at Ballet Theatre became increasingly unpleasant. Between *Undertow* and his remarkably different inspiration, *Shadow of the Wind,* the sensitive Tudor suffered humiliation and personal disparagement from Ballet Theatre's management. There is a series of still extant letters between Tudor and Lucia Chase and her lawyers and business managers from July 1945 through the beginning of August 1945. They document the bitter quarrel that sealed his uncertain fate at Ballet Theatre and prophesied his eventual departure.

What may have triggered the explosive conflict was Tudor's work on those two Broadway shows in 1945. Tudor was refused a leave of absence to work on the second one, which shocked him, as he thought the management would release him briefly since he was not needed by the company.

Tudor wrote to Lucia Chase on July 13, 1945:

> I find it very sad to have to write you this letter after such a long and for the most part, close association with you. But in view of the past month's developments. . . . It is in friendship and loyalty to you that I chose to write you and do not engage attorneys to protect my interest and redress damages which are far more important to me than hard cash. . . . I feel that such damage and injury has been done to me as cannot be repaired. . . . I am ready to continue with the company until the end of this present period August 31, 1945 on condition that I shall be fully released after this date. If this condition is not met, I must take this matter to my attorneys with instructions that they terminate my agreement for the breaches by B.T.

The letters flew back and forth, getting more and more serious, closer to threats of quitting, dismissal, and suits. What was it that drove Tudor to such fury? In the past, he found that the management preferred their Russian émigré dancers and choreographers, offering them better billing, more credits and opportunities. His ego had been wounded; his pride hurt.

In addition, Tudor's fate was inextricably linked to Hugh Laing's, and, if Hugh was slighted, then Tudor responded personally. In a letter to Chase, also on July 13, Hugh explained that his pride and his reputation had been severely abused and ignored by several management oversights. "I hate to do this but I know that it is essential to my integrity." Apparently, in the publicity, Anton Dolin's name was placed above Laing's and was in larger type. Dolin, though an Englishman, was originally a member of Diaghilev's Ballets Russes.

Laing and Tudor were eventually able to work out some of these problems, and they again resumed work at Ballet Theatre. But as the letters demostrate, Tudor and Hugh were wary employees, and these experiences paved the way for their departure in 1950. In a sad and brief letter sent before the fall season in 1945, Tudor says, "I hereby consent to the use of my picture as a dancer in Ballet Theatre's Souvenir program for the 1945–46 season provided that the words 'On Leave of Absence' appear below or to the side of this picture.

Several of these letters allude to one of the major disappointments in Tudor's career just before the first Ballet Theatre European tour to England, a very symbolic trip for Tudor, in 1946. The Artistic Advisory Committee of Ballet Theatre wanted Tudor to be the artistic director of the company. They were a group who counseled the board and the directors of Ballet Theatre. The Committee suggested ideas for ballets and discussed repertoire and other issues that concerned the company. At this period it consisted of Antony Tudor, Agnes de Mille, Aaron Copland, Jerome Robbins, Lucia Chase, Oliver Smith, John van Druten, and Henry Clifford. The Committee voted that he assume that prestigious role just before the tour to London, but Lucia Chase and Oliver Smith were dead-set against it and changed his title to "Artistic Administrator," which was certainly not the same thing.

He remained as artistic administrator for less than two years, as he was no longer listed as such in the 1948 program. Probably the most significant omen of Tudor's worsening situation at Ballet Theatre was the fact that after 1947, fewer of his ballets were maintained by the company.

But in 1946, as Ballet Theatre's new artistic administrator, Tudor had apparently smoothed over his quarrel with Lucia Chase. As one of his first administrative decisions, that year he invited Margaret Craske to become the ballet mistress of Ballet Theatre. Tudor had started as a student of Craske's and later found her an outstanding and dependable colleague. When Ballet Theatre and the Metropolitan Opera House collaborated to open a school at the Metropolitan in 1950, she remained there to teach. It later became the Metropolitan Opera Ballet School, and Craske assumed the role of co-director during her nearly twenty-year association with the school.

Tudor returned to London on July 4, 1946, on tour with Ballet Theatre. It was an emotional journey, as it was his first glimpse of London since 1939. Tudor remembered the experience as rather painful. After all, he had not rushed back during the war to share his colleagues' hardships. The British neither forgot, nor, perhaps, forgave.

Apparently Ninette de Valois, now Artistic Director of Sadler's Wells, had arranged tickets for Tudor and Laing to attend a performance while they were visiting. Despite the complimentary tickets, for a while Tudor and Laing thought they were despised: "Ninette organized a situation where, for a while, absolutely no one would speak to me. Blackballed! And so we went to the bar during intermission and Ninette was coming down the staircase and she ran down to embrace me. And that was the end of that."[4] Those must have been chilling moments! Many years would pass before Tudor was really welcomed in London.

London's response to Tudor's ballets varied. They applauded *Pillar of Fire:*

> "It is a work of great beauty and sincerity, in which Dance as the expression of the most complex emotions is allowed to tell the story." The critic praised Tudor for "preserving his belief in the value of Ballet as a serious contributor to the Art of Theatre in a country, where, as far as I can judge on the present showing, the vast majority of ballet audiences consider it as a joke to be enjoyed lightly in revue-like fashion."[5]

Tudor's *Undertow* fared less well, "being linked with the current Hollywood craze for the psychological film."[6]

After Ballet Theatre returned to America, Tudor spent many months with them on tour: from Boston, to San Francisco at the New Year, to Los Angeles, to Chicago in April, and, with Alicia Alonso's help, to Havana in May. The year's tour (in 1947) was broken by a season at the New York City Center from November 19 to December 17. The second New York season of that year, this time at the Metropolitan Opera House, extended from April 4 to May 8. As artistic administrator, Tudor had many responsibilities in addition to keeping his own ballets in good condition. That season they included *Dark Elegies, Gala Performance, Jardin aux Lilas, Pillar of Fire,* and *Undertow.* There was no new Tudor ballet.

On January 13, 1948, Tudor reworked and restaged *Time Table* for the New York City Ballet, a ballet originally created in 1941. Tudor's growing disenchantment with Ballet Theatre motivated him to explore work with the New York City Ballet in the hope that they would recognize and appreciate his talents. Perhaps this brief encounter with the New York City Ballet encouraged him to seek opportunities elsewhere.

Meanwhile, Ballet Theatre's financial situation became increasingly more desperate. They began to threaten that they would not be able to sustain their fall 1948 performance season.

While Ballet Theatre was in crisis, Tudor was making plans to create his next major ballet. For some time Tudor had wanted to produce a full evening ballet based on Proust's *À la recherche du temps perdu*. He even visited Paris to discuss the ballet with Proust's relatives. Though Tudor announced to Isolde Chapin in *Dance Magazine* that his next important work would be the Proust piece, he set the Proust aside and created a new work, *Shadow of the Wind*, which turned into the big ballet he had been promising.[7] The rehearsals for *Shadow* added up to an extravagant 358 hours, and a great deal of money was invested in elaborate sets and costumes.[8]

Shadow of the Wind (April 14, 1948)

Shadow of the Wind is based on Gustav Mahler's score for *Das Lied von der Erde* (*Song of the Earth*), a symphony for contralto and tenor. Mahler chose six poems based on ancient Chinese texts that interpret various theories of human existence. Tudor explained that "his ballet symbolizes the impermanence of existence, the Chinese philosophy of accepting the mutations of life and bowing before them. Like the seasons, human experience is cyclical and has no sudden beginning or end."[9]

Mahler's six poems are titled in English: I. The Drinking Song of Earth's Sorrow (for Tudor's "Six Idlers of the Bamboo Valley"); II. Autumn Loneliness (for Tudor's "The Abandoned Wife"); III. Youth (for Tudor's "My Lord Summons Me"); IV. Beauty (for Tudor's "The Lotus Gatherers"); V. . . . Wine in Spring (for Tudor's "Conversation with Winepot and Bird"); VI. The Farewell (for Tudor's "Poem of the Guitar").

Shadow of the Wind was Tudor's first ballet with an Eastern theme since *Atalanta of the East,* his light foray into Orientalia, in 1933. Tudor's affection for Mahler's music, which he previously used in *Dark Elegies,* was perhaps misplaced in this case, as a number of critics believed the music antithetical to Tudor's themes. The large orchestra gave the work a sumptuous and resonant sound, and the length of the ballet, more than an hour and probably the longest of all of Tudor's works, make one think that this is indeed a major choreographic piece.

To the first song for solo tenor, "The Drinking Song of Earth's Sorrow," five men, one an old poet, another an old warrior, dance to the words, "O man, what is the span of thy life? / Not a hundred years art thou permitted to enjoy / The vanities of this earth!" The second song, for alto, laments

Shadow of the Wind, *1948. Nana Gollner in "Poem of the Guitar."* (Photo by Carl Van Vechten)

loneliness in autumn and the passing of life, and is transposed in dance terms as a solo for an abandoned wife: "Autumn in my heart too long is lasting, / O Sun of love, never again wilt thou shine to dry my bitter tears?" The tenor's song of youth describes a pretty Chinese landscape with a green and white porcelain pavilion in a little pool that is reached by a bridge of jade. "My Lord Summons Me" is the name of a dance for a girl, a boy, and the ensemble. "The Lotus Gatherers" is the title of the next section, which is performed by a girl and the ensemble to a contralto solo about beauty, where youthful maidens pluck lotus flowers at the shore, where lovers play at longing. The fifth part, "Conversation with Winepot and Bird," is accompanied by the tenor lament that all life is woeful and wine the only reality. The final dance, "Poem of the Guitar," is performed to Mahler's "Farewell" for contralto: the poet waits for his friend for a last farewell but looks to the coming spring to waken the world anew.[10]

Zachary Solov remembered that Alicia Alonso's solo of the Abandoned Wife deeply touched him while he was standing in the wings watching the ballet. "I loved the moment when Alicia was on her knees sitting. There was no dance movement. She cried and her tears dropped into the lotus bowl. She picked up a tear and looked at it, and let it fall back down with a slight oriental movement of the arm." Solov commented that Tudor barely rehearsed for *Shadow of the Wind,* despite the number of clock hours: "We would arrive in some god-forsaken little town on the road, go to the local Elks Club and wait, while Tudor stood there and contemplated. This went on for the whole tour. And yet the costumes and the scenery were so opulent, so magnificent, so perfectly designed, by comparison with the choreography."[11]

Robert Sabin, in *Musical America,* revealed that the scene designs, executed in the Chinese manner, "set a record for lavishness. At times Tudor follows almost literally the Chinese poems translated by Hans Bethge and used by Mahler. In other episodes, Tudor imposed original dramatic action upon the emotional situation portrayed in the text and music." Sabin spoke about the ballet as having striking paradoxes: the poems are written by a man and speak from a man's perspective; Sabin, however, understood that Tudor composed the dance from a "feminine point of view," as he gave the most stirring and "convincing choreography" to women dancers. Sabin preferred the scene in which Nana Gollner does a solo, "Poem of the Guitar." "Here again the poet's objective vision of the eternal challenge of human isolation and the reassuring fertility of the earth is transformed into a subjective situation."

Sabin disliked intensely the segment "Conversation with Winepot and Bird," as "Mahler's vision of the exquisite coming of spring was reshaped and diminished into a comedy where a huge bird apparently the result of the mismating of a parrot and a pelican, was dragged across the stage on wires. Was Tudor lured by the spectacular and 'gorgeous' scenery that Mielziner produced for this ballet and did he thus neglect the 'psychological penetration' of his characters?" Sabin also criticized the long sleeve-dance with the corps de ballet, where the "heavy tread of the music evokes the most heartbreaking despair." Perhaps the most whiplike thrash that Sabin aimed at Tudor was the comment that the choreographer "borrowed movement from previous works."[12] Even Sabin's praises became critical, he wrote that the dancers did their best with "weak and uninventive" dances.

In *Dance Observer*, Joan Brodie testified that the mixing of styles doomed the work to failure: "It is impossible to correlate the visual and the auditory at such opposite poles. Perhaps it is the underlying confusion of styles . . . when we try to combine the Chinese, German, and other inspirations with the satin toe slippers and the classical movements."[13]

In another negative assessment, Walter Terry faulted Tudor for not paying attention to the lyrics of Mahler's songs, referring rather to the Chinese poems of Li Po. Tudor "became lost" as the dances did not reflect the music, nor did they represent an independent exploration of the rhythms and shapes in the poetry. The ballet moved within a mood rather than a dramatic structure. Terry noted that Tudor was at his best when he used the ballet technique with "freely expressional movements of modern dance." But he added certain movement idioms of Oriental dance, and "these actions appear as so much adhesive tape on the ballet-modern frame." In the same way, Terry continued, gesture was used for dramatic implication as well as ceremonial effect; long sleeves were manipulated in rhythmic patterns; and certain steps suggested the acting style of the Oriental theater. "At its most effective, orientally speaking, it is mid-Denishawn in period." Terry went on to say that there were moments when Tudor gave the dancers an entirely Oriental feel and pantomime, which afforded a pleasant and decorative vision. The blame might rest with the dancers, Terry thought, who seemed ill at ease with the choreography. Terry praised both Muriel Bentley and Nana Gollner, as they were given movements of emotional force.[14]

John Martin echoed Terry's remarks: "What happened on the stage was on the surface a considerable disappointment. . . . The work demanded great style and inward poise from its dancers." Martin found the ballet thoroughly unconventional and far removed even from Tudor's own repertory.[15]

The distinguished drama critic George Freedley wrote in the *Morning Telegraph* of New York:

The beauty and purity of austere Chinese art was successfully evoked and lighting was particularly effective. We can all share Nora Kaye's disappointment that her illness kept her from dancing in the new opus and Nana Gollner, a lovely and capable substitutue, made her appearance on the last, the 6th poem with Hugh Laing and Dimitri Romanoff. It is difficult to describe this ballet because it is all so much a question of mood.[16]

Walter Terry in his Sunday column further considered his original and disturbed response to *Shadow of the Wind,* the first American ballet Tudor had choreographed in three years. Terry cited some of Tudor's distinguished past creations in order to explain why he was shocked by this work. He repeated that it was disconcerting to watch the ballet without being able to make sense of Mahler's lyrics:

> While the words are impressing specific images and moods upon the mind, the stage is concerning itself with rarely related actions, with movements which are laxly pictorial in a pseudo-Oriental sense and which all too infrequently take on the ballet-expressional dance cast which would lend itself to an immediate relationship to the music at hand.[17]

It seemed that many people felt that the ballet would have worked had Tudor selected a single style for his movement choices. Several dancers have testified to the beauty and value of this unusual work, and ballet historians have lamented its complete loss. For example, in conversations with Fernau Hall, Margaret Craske maintained that it was a jewel of a choreographic work and that Alicia Alonso's variation was most moving.[18]

Critical comments must have had a major influence on the decision of the management to cancel *Shadow* after only six performances. Another important reason for the cancellation of further performances at the time may have been (according to Fernau Hall) that the Mahler estate demanded an orchestra of 106 and Lucia did not want to pay the very expensive orchestra fees considering the length of the ballet.

In an angry and vituperative mood, Fernau Hall accused Chase: "That was the most terrible crime that Lucia ever committed, to destroy that ballet!"[19] It was never performed again.

Besides the devastating problems with *Shadow,* other parts of Tudor's life seemed to be coming apart. In addition to his feelings of insecurity and distress at Ballet Theatre, his close relationship to Hugh Laing had been severely strained when Hugh married Diana Adams in 1947. Adams was a beautiful and talented dancer twenty-one years old, while Laing was fifteen years her senior. Tudor also pushed away his old friend Agnes de Mille as a result of one of his cruel and nasty barbs. When de Mille's *Fall River Legend* opened on April 22, 1948, to good reviews, eight days after the disastrous

Agnes de Mille and Antony Tudor in a serious conversation, with ballet student in background, 1956. (From the Performing Arts Research Library)

opening and closing of *Shadow,* Tudor claimed that de Mille stole her idea for *Fall River Legend* from his ballet *Pillar of Fire.* She was understandably furious and refused to speak to him for several years.[20] This was neither the first nor the last pernicious remark that Tudor made to de Mille. Unfortunately, Tudor had a diabolical and sardonic side that alienated many of his friends and dancers.

In the summer of 1948, faced with rising costs of production, Ballet Theatre was forced to cancel plans for an autumn New York season in order to dedicate that time to raising money. The following spring, the company was able to undertake a tour of the East and Middle West from March 19 to April 9 and to fill an engagement at the Metropolitan Opera House from April 17 to May 8. "A year of not working together cannot benefit a company, and it did not benefit this one. . . . Troubles continued to plague Ballet Theatre as they dispensed with an autumn season in New York in 1949 and, instead, did a national tour from November to March 1950."[21]

A little more than a year after *Shadow of the Wind,* Tudor presented a modest little ballet at Jacob's Pillow.

Margaret Craske and Antony Tudor. (Courtesy Donald Mahler)

The Dear Departed (July 15, 1949)

When Tudor returned in 1949 to the relaxed summer atmosphere of Jacob's Pillow, he offered ballet students and several of his dancers the chance to perform some of his best works, including *Gala Peformance, Jardin aux Lilas,* and *Judgment of Paris.* In addition, he created a curious little work, *The Dear Departed,* which was about a lady who squeezed out of her bottle and refused to return, much to the dismay of a wooing Chinese man. It is a wonder that Tudor chose another Oriental theme after the mild catastrophe of *Shadow of the Wind.* Was he healing his wounds? Though we do not know the source of this ballet caprice, it is probably one of the poems that inspired *Shadow of the Wind.* In any case, it offered Hugh Laing and Diana Adams a vehicle for their marvelous talents.

A brief, poorly lit film clip gives a fleeting but sure impression of the piece. Laing is dressed as a Chinese man with a beard, a headband, and a loose, pajama-like costume. Diana Adams is on pointe in a gauzy skirt with bodice and pants. Diana Adams called it a "sweet little pas de deux. I was a genie who came out of a bottle."[22] She gave the appearance of being filmy and wraithlike as she floated tantalizingly near Laing, having just alighted from a small vessel nearby. As Laing lifted her, she pointed to the bottle, moving away from him in *bourrées* and little, flitty steps.

Walter Terry explained that the ballet was a new work with variations on three classic gestures:

> Mr. T's little legend was charmingly danced by Diana Adams and Hugh Laing. Through the accuracy of their gestures, through their knowing use of accent, the two made manifest the intrinsic wit of this balletic trifle. Particularly winning was Miss Adams's exit when she coyly refused to return to her bottle and in making her sinuous and curiously inviting escape, became *The Dear Departed.*[23]

Wayne Smith called it "a light work, full of comedy. It tells the fanciful story of an old Chinese sage? Philosopher? at any rate an immoderate imbiber who finding his bottle empty prays that it be filled. In answer to his prayer, a genie appears in a burst of flame and he promptly attempts to woo her and inveigle her into the bottle, but without success. Hugh Laing did a commendable job as the old Chinese man and had great opportunity to show off his ability as a dramatic dancer. Miss Adams made a piquant and beautiful genie." In Smith's interview with Tudor, Tudor modestly remarked that the ballet "was hastily thrown together."[24] Smith later wrote, "Hugh showed exceptional ability as a dramatic dancer and did a commendable job with his restraint in those phrases that burlesqued the classical pas de deux giving them a subtle rather than a broad touch of humor, while Miss Adams was a beautiful and alluring genie."[25]

The Dear Departed has never been restaged.

In an unexpected and dramatic move, Tudor accepted an invitation to the Royal Opera House in Sweden to assume the artistic directorship of the Royal Swedish Ballet. The summary defeat of *Shadow of the Wind,* plus the serious financial problems at Ballet Theatre, undoubtedly influenced Tudor's decision to go to the Royal Opera House in Sweden in 1948 for the 1949–1950 season as artistic director, a position that he had always wanted and was never granted at Ballet Theatre.

Tudor spoke of his sojourn in Sweden and the Swedish dancers with Jennie Schulman in *Dance Observer.* He remembered: "The dancers were far from strong, but they had great performance value in that they were trained for the stage all their lives. They never fear a lack of finances. However their interpretive powers were limited when it came to ballets like *Jardin,* since they are at their best displaying emotions and couldn't convey emotions withheld.[26] He reiterated to Marilyn Hunt that they "all grow up in the school together. It's like one understanding family. At Ballet Theatre, you never know who's going to come in next week, or who's going out."[27]

Why did Tudor like Stockholm? First, for its sense of architectural space. The city is surrounded by water, canals and sea. It is situated on

Antony Tudor and Mariane Orlando with the Royal Swedish Ballet rehearsing Pillar of Fire, *1962, in Stockholm.* (Courtesy of Mariane Orlando)

fourteen islands that have easy access by boat. There is always a distance to look into where one's soul can wander at ease. The buildings tend to be large and imposing, not fanciful and filigreed. They have an austerity and clarity of line that appealed to Tudor, and, as anyone may notice, the city is very clean; immaculate, by New York standards.

Tudor had visited Stockholm in 1948 at the invitation of the Royal Swedish Opera director, Joel Berglund. He discovered an opera-ballet company of approximately forty dancers that had not kept up with other companies, probably as a result of the war. Their technique was a bit ragged, and their style had a prewar aura. Tudor had heard of Constantin and Nina Koslovsky, a Lithuanian couple recently defected to Stockholm who briefly trained in Moscow. He hired them to be ballet master and mistress to put the company in shape. They subsequently remained as teachers in the school for many years. Refreshingly open and direct, several of the Swedish dancers later revealed Tudor's nasty cracks about their woolly feet and evident lack of training. No one knew better than the Swedes what their technical weaknesses were.

The Royal Swedish Opera House, a beautiful structure with typical red and gilt interior decorations, holds an audience of 1,150 people. Though the theater is remarkably equipped, it continues the old tradition of a raked or sloped stage, and one of the large dance studios where the company rehearses is also (mercifully) raked.

During the 1949 fall season in Stockholm, Tudor mounted *Jardin aux Lilas* and *Gala Performance,* and he also restaged *Petrouchka,* which had been re-created by Ballet Theatre in 1948. The following spring, after restaging *Giselle* (February 9, 1950) for the Swedes, Tudor travelled to Paris and then to Milan.

The next year he was replaced by Mary Skeaping, a close English colleague, as Tudor decided to return to New York and to fulfill new obligations to the Juilliard School and the Metropolitan Opera Ballet. Skeaping became director of the Royal Swedish Ballet and held that position from 1953 to 1962.

As America had become the center of dance activity during and after the war, perhaps Tudor felt he had to return to New York so as not to be forgotten by a fickle entertainment world. One might also imagine that the long, dark days of Sweden's winter had a negative effect on Tudor's already variable moods.

Tudor had also been slated to produce a ballet for the dramatization of James Joyce's *Finnegans Wake.* In the July 1949 *Dance News,* there was an extensive article on Tudor's plans for the piece. Since there were no reviews of or program for the musical play, it is likely that the production never occurred. But what a wonderful idea it would have been!

Voyaging by ship, Tudor made his way back to New York to choreograph what turned out to be a minor piece for Ballet Theatre, *Nimbus,* a work that he developed from Louis Gruenberg's Violin Concerto. One wonders why, after all that passed between him and the Ballet Theatre management, he returned to the lion's den.

Nimbus (May 3, 1950)

Nimbus represented a departure from Tudor's earlier presentations in New York. It was a light situational work in mood, energy, and emotional range, not unlike a musical comedy dance, as were *Goya Pastoral* and *Time Table.*

The program notes read: "A working girl dreams of a boy from next door. Her dream-self, alluring and beautiful, has no difficulties attracting his attention. The girl wakes . . . it is a hot night so she goes to the roof of her house where among many 'night birds' she finds her dream beau as well."

The definition of "nimbus" in classical mythology is a shining cloud

Nimbus, *1950. Hugh Laing as the Dream-Beau.* (Photo by George Platt Lynes)

sometimes surrounding a deity when on earth. In this case, the fairy goddess who haunts New York roofs is the sleeper's alluring dream, and her halo follows her every move.

For the opening of the ballet the viewer discovers Nora Kaye lying in a huge, stylized bed, fitfully dreaming. The bed sits almost perpendicularly on the stage so that one clearly looks down upon the sleeping girl. The initial scene presents the dance relationship between the sleeper and her dream-self. Her imaginary idealization of herself (Diana Adams) meets a handsome stranger (Hugh Laing) and captivates him with her gentle and languorous qualities. Her movements also reveal a sense of satire as the sleeper envisions herself as a "voluptuous siren" luring her willing victim into the mazes of the wicked tango, as she entices him with her own sweetly innocent ideas of a sensuous dance. The sleeper's dreams take place at the foot of the bed, occasionally with ecstasy, and disgruntlement and her own tossing actions mirror the full-motion dream that transpires around her. When the girl awakens, she restlessly tries to recapture the fragments of her shattered dream. Unsuccessful, she climbs to the roof of the apartment, where the second scene emerges from the turntable set.

The roof is the refuge of other sleepless city dwellers, such as someone seeking a cool breeze and unable to find a perfect place for the mattress, in addition to a lady enchantress in her kimono and curl-papers, and a contrasting woman figure given to maudlin drunk states—shades of *Judgment of Paris.* There on the rooftop, the sleeper encounters the actual man of her dreams. The rest of the ballet is a long *pas de deux* occasionally interrupted by the passage of other "Night birds," in which the girl and the man discover the beginnings of a shy and hesitant love. The ballet ends as the girl returns to her room, almost walking on clouds.

In Jennie Schulman's article in *Dance Observer,* Tudor remembered *Nimbus* as having a cool reception. "It started out great, but I never was one for doing a ballet in three weeks and that's all the time allotted me. It was money trouble as usual. . . . I still have plans for reworking it satisfactorily."[28] The opening night of *Nimbus* was greeted warmly with many curtain calls, according to Walter Terry's review the next day. He alleged that, when it was good, it was terrific, but it suffered from choreographic sinking spells. Terry effusively noted that the dancers acquitted themselves extremely well, "and when the material was meaty, they danced in an exciting fashion. Nora Kaye was particularly fine as the dreamer. Her awakening scene was worth the price of admission so perfectly was it done, with just the right touches of beauty of action and gaucherie."[29]

John Martin wrote: "Tudor has turned away from his psychological explorations and given us a completely uninvolved, fairly sentimental little tale of a lower-middle class romance—Mr. Tudor makes a strong case for apartment house roofs; you meet such interesting people there." Martin described the lyricism and "physical eloquence" of Nora's dancing and praised Diana Adams for her long opening sequence, which explored her "lyric potentialities and the strong, clean technical force that underlies them with extraordinary beauty and touches of wit and warmth. It seems that all

Hugh does is lift, lift, lift. The scenery by Oliver Smith is exceedingly ingenious.[30]

Unfortunately *Nimbus* has never been restaged.

Hugh Laing and Diana Adams left Ballet Theatre for the New York City Ballet in 1950. The story of their departure seems to vary, depending upon whose point of view one hears, and certainly it had an enormous effect on Tudor. Apparently Laing's real reasons for leaving Ballet Theatre were simply that Lucia Chase would not promote Diana Adams to principal soloist, and that Hugh was not being paid as high a salary as John Kriza. Later that year, Nora Kaye also left Ballet Theatre and joined New York City Ballet, as the choreographers she wanted to work with, Tudor and Robbins, were both there. Robbins joined New York City Ballet as associate artistic director and dancer in 1949.

Though Tudor did not create any more new ballets for Ballet Theatre until 1975, he returned there often to oversee rehearsals and restagings of his earlier ballets; he became something of a rusty icon, prayed to at convenient times. Eventually, Hugh Laing resumed work at Ballet Theatre; Nora Kaye danced with New York City Ballet until 1954; while Diana Adams remained at New York City Ballet until her retirement, becoming one of Balanchine's favorite ballerinas.

CHAPTER SEVEN

Tudor the Educator

*I*t is at this point in Tudor's career that his activities seem to multiply and unfold with astonishing speed. His rupture with Ballet Theatre left him free to accept many opportunities that led in different directions, sometimes all over the map. He created several works for the New York City Ballet; he made little pieces for the Juilliard students and the Jacob's Pillow dancers;[1] he taught at Juilliard and the Metropolitan Opera Ballet School and gave lectures in the community on dance theory and history. He continued to promote and restage his better ballets, but no new major work arose during the long period from 1950 to 1963. How can one explain this lull? But then again, there is also the question of what makes a big ballet, a piece worthy of an illustrious ballet company and dance historians' and critics' commendations through the centuries.

The chamber ballets that Tudor created for Juilliard students and Jacob's Pillow dancers, both professional and student, during these years were important and innovative explorations of the ballet technique and scenarios. Unfortunately, many of these ballets are completely lost, as there is no longer anyone who can reconstruct them,[2] although there are contemporary accounts of them. Fortunately, some of his less famous ballets have been notated and have received some fine and faithful revivals by regional companies since Tudor's death. Only time will tell, but why should not American Ballet Theatre or the New York City Ballet restage *Little Improvisations* or *Offenbach in the Underworld?*

After many weeks of traveling in Europe during the summer of 1950, Tudor spent five weeks in France, where he visited Serge Lifar and the Paris Opera Ballet. Unhappily, no performances of his ballets in Paris developed from this trip. During the same summer, Ballet Theatre, now called the American National Ballet Theatre, had its grand tour to Europe, playing at Covent Garden in England; the Opéra in Paris; and Germany, Scotland, Italy, and Switzerland.

After the rather abortive stint with the Royal Swedish Ballet, Tudor began his long association with the Metropolitan Opera in the fall of 1950. *Opera News* announced that Tudor had been hired as the administrative director of the Metropolitan Opera Ballet Company and the Metropolitan Opera Ballet School.[3] He remained the director of the Met ballet company until 1962, and ran the school until 1966. In the beginning, Ballet Theatre dancers were used routinely in the operas that needed soloists. Early in 1950, Ballet Theatre signed a contract with Rudolph Bing, the general manager of the Metropolitan Opera. The Metropolitan's corps of dancers was retained, with a considerable number of replacements being made. Ballet Theatre's Nana Gollner was signed as *première danseuse,* and Tudor was named choreographer and ballet master as well as administrative director. The association with Ballet Theatre endured only for the season.

In his first opera choreography in the United States, Tudor presented three dances for Verdi's *La Traviata,* two dances for Gounod's *Faust,* and the waltzes for *Die Fledermaus.*

Tudor for once was creating quickly. Perhaps coincidentally, the reviews were not very favorable, and in 1951, Tudor relinquished his position as choreographer and ballet master at the Met. Zachary Solov, who had a talent for the quick, exciting, and limited needs of opera ballet, took over his position. It was evidently a mutually agreeable change of affairs, as Tudor remained as administrative director of the Metropolitan Opera Ballet Company for another eleven years.

No doubt Tudor was pleased enough to give up making the little opera ballets, since he was busy choreographing for the New York City Ballet, producing two new full-length works for them in fairly short order. The first was presented just two months after his underappreciated dances at the Met.

Lady of the Camellias (February 28, 1951)

Program notes for the ballet:

> Scene 1. Paris 1848, Party "Chez Prudence"
> Prudence, Vida Brown
> Guests, 4 women and 5 men
> Scene 2. In the country, 2 months later
> Marguerite   Diana Adams
> Armand   Hugh Laing
> M. Duval   John Earle
> Scene 3. Paris several months later
> Scene 4. Marguerite's bedroom

Lady of the Camellias, *1951. Hugh Laing and Diana Adams.* (Photo by George Platt Lynes)

When Hugh Laing, Nora Kaye, and Diana Adams fled to the New York City Ballet during the summer of 1950, they were warmly welcomed by Lincoln Kirstein and eventually given some challenging roles. At the same time, Tudor was invited by Kirstein to create works for his "own" dancers, Diana Adams and Hugh Laing. According to Selma Jeanne Cohen, "The repertory of the New York City Ballet, consisting chiefly of abstract works by George Balanchine, was weak in the area of dramatic dance, where these artists excelled, and it was logical that Tudor should be asked to provide such vehicles for their talents as Balanchine was not interested in supplying."[4] Comfortable with dramatic themes and with theatrical stylization, Tudor appreciated this interesting challenge. Lavish costumes and luxuriant scenery originally created by Cecil Beaton for the Ballet Russe production of *Camille* were put to use once again in Tudor's *Lady of the Camellias.*

The novel (1848) *La Dame aux camélias* by Alexandre Dumas *fils,* which inspired this ballet, has been transformed successfully into operas, plays, and movies. The story concerns Marguerite Gautier, a Parisian courtesan, called "la dame aux camélias" for her passion for this flower. She falls deeply in love with Armand Duval. The two live in quiet happiness in the country until Armand's father visits Marguerite, determined to separate his son from the "fallen woman." When bribes fail, the father implores her to give Armand up for his sake. Convinced that the love affair is damaging to Armand, she returns to Paris, letting him believe she has left him for a rich nobleman. But the sacrifice has been too great, and her already consumptive condition grows rapidly worse. At first angry at the seeming betrayal, Armand eventually learns the truth and hastens to Paris in time for Marguerite to die in his arms.

According to Lillian Moore writing in *The Dancing Times,* Tudor typically did not probe new depths of the subconscious but rather interpreted a familiar and well-loved story.[5] The four scenes of this ballet are the same as the four scenes of *La Traviata,* the Verdi opera based on the same theme.

In the opening scene, a splendid solo for Marguerite demonstrates her flirtatious ability to charm; she audaciously denies Armand's entreaties: "His protestations of devotion are the theme of another admirable passage. Her awakening interest in him, the jealousy of her protector the Count, her fragility and illness are all beautifully and intricately woven into dance design."[6]

In the only segment of film from the ballet, the *pas de deux* between Armand and Marguerite emphasizes their intense attraction to one another. Her costume accentuates her fragility; she wears a light, chiffon dress with a peplum, lacy, with a low neckline. They both have billowing, soft sleeves that shadow their interconnecting arms. Laing and Adams enter right upstage diagonal. She moves into arabesque, reaching towards him in a deep lunge; she flies through many little, wispy steps with fluid, flowing arms. Often they hold one another as if whirling into a nineteenth-century ballroom dance. She sustains her coquettish qualities while Armand plays the fervent lover. When the dancers face each other, they do *glissades* or quick shifts of movement in all directions, giving the duet an ambivalent, occasionally frenetic tone. They continuously move indirectly and playfully. The partnering is very vertical, again with a sense of mid–nineteenth-century formal dancing. Marguerite's torso is almost always straight with some arabesque balances where she changes hands (as in "Rose Adagio"). As the *pas de deux* concludes, their dancing becomes closer, building gradually in intensity. She finishes on the floor, kneeling; they both face the audience while she looks up at him.

Their romantic *pas de deux* is followed abruptly by a ballet pantomime scene in which Armand's father forces Marguerite to promise to leave Armand. During the third scene, which takes place in a spacious salon, the surrounding noisy gambling activity makes the meeting between Armand and Marguerite all the more intense. A powerful moment occurs when Armand gambles against the Count, wins, and, unlike the opera, where Armand showers Marguerite with gold, he offers her his winnings with a stiff and formal bow. The final scene in Marguerite's bedroom recaptures the intimacy of the pastoral *pas de deux,* although with more pathos. Marguerite's dying moments, realistically painful ones, mirror the suffering her love for him has caused her. Lillian Moore described the last movements, "where pain seems to wrench her from Armand's arms to fall face downward on the bed, in a shuddering, stiffening heap."[7] Tudor was praised for wisely choosing musical excerpts from several of Verdi's operas that provided the ballet with nuances of subdued tenderness and gave it an independence it would not have possessed had he used the music for *La Traviata*.

Walter Terry reminded his readers that this was the third attempt in recent years to translate the story of Camille into the language of dance. "The Tudor work . . . was curiously uneven." The opening scene with couples at a party presented Marguerite and Armand responding to one another in a "surprisingly casual" way. "However, the love duet later in the country scene brings Tudor's familiar use of movement to penetrate feelings of the heart. Gestures of tenderness, some gently sweet and others playful, were mated with larger actions indicative of soaring happiness, of surging love.

Terry faults Tudor for the ensemble work, finding it spare and unimaginative. He also found the father of Armand a negative feature on stage, a personality who was ineptly conceived and did not offer the plot significant interest. "However, the death scene, with its faltering solo and its brief and sad duet communicated strong emotional values. Why the dancers surrounding the main characters filled and perhaps drained the stage of its energy could only be due to a lack of rehearsals, and perhaps the ineptitude of New York City Ballet dancers to find a dramatic reason to be on stage." Apparently Tudor made an unannounced appearance as the father of Armand, and Terry found his acting stilted and severe. In contrast, the performances of Diana Adams and Hugh Laing were admired for their remarkable refinement. In the audience, Greta Garbo, another Camille, applauded the ballet heartily.[8]

John Martin's appreciation was more sympathetic and positive. On the evening that Martin attended, John Earle, "who bears an uncanny resemblance to Tudor," played the father. Tudor's method of choreography "was inspired, rather like the eighteenth century choreographer, Noverre, who developed the invention of a choreographed drama, a pantomimic spectacle where gesture imparts meaning." Martin continued, "It contains a minimum of pure dance; its action unfolds in terms of what might be called abstract dialogue, varying from a hint of the formal miming of the nineteenth-century ballet, through fairly natural gesture to a heightened expressiveness towards inward feeling when the situation touches deeper emotional levels."

As Martin announced, Tudor once again created a new way of presenting a ballet, accomplished with originality and sensitivity of texture:

He understood theatre, perhaps better than any American choreographer except Robbins, who learned a great deal from Tudor. He understood the stultifying qualities of nineteenth century values that the novel exposed and succeeded in reinforcing these facts with his exposition of the love story, their tenderness, anger and sentimental misunderstandings always haunted by the over-hanging shadow of her death. During the pas de deux which all the critics acknowledge is the high point of the ballet, the theme of Marguerite's illness is given subtle expression.[9]

Lady of the Camellias had no revivals or restagings.

The following spring, in April 1951, Tudor began a series of lectures, "Ballet Today," at the Henry Street Settlement Playhouse. Tudor gave and received much pleasure from his Toynbee Hall and Morley College discussions and his abiding interest in educating the public about ballet. His listeners learned that Tudor's respect for pure movement remained paramount in his choreographic process when he asserted that dance should develop from dance or movement, not from music or literary ideas.[10]

It is worth remarking that Ballet Theatre did not fare very well without the presence of Tudor and his dancers. Lillian Moore pointed out that "Ballet Theatre has suffered noticeably from the departures of Kaye, Adams, and Laing, and especially from the absence of many of the great Tudor ballets."[11]

Just four months after *Lady of the Camellias,* Tudor unveiled another new ballet, *Les Mains Gauches,* to music by Jacques Ibert, this one choreographed for the students and professional dancers at Jacob's Pillow. At this summer dance theater, along with Craske, Tudor continued to give ballet classes and to work on ballets that interested him.

Les Mains Gauches (July 20, 1951)

His first ballet at the Pillow, *The Dear Departed* (1949), was a sweet, slight work; *Les Mains Gauches,* also quite short (eleven and a half minutes), was considered a "merry prank," "a witty comment on customs and superstitions associated with the left hand."[12] During the same summer, Tudor decided to return to one of his most challenging ballets and to re-create *Dark Elegies.*

Walter Terry observed:

> Sallie Wilson, Zebra Nevins and Marc Hertsens did a fine job of projecting the work's pleasantly sinister atmosphere. There is no literary line to *Les Mains Gauches,* but incident, conflict and mood give emotional colour and dramatic force to a dance creation which plays wittily and teasingly upon customs,

instincts, mysteries and patterns of action associated with the left hand, with things sinister.[13]

In a rather fuzzy film of the ballet from Jacob's Pillow, Zebra Nevins and Marc Hertsens play two rather lost souls while Sallie Wilson dances the role of Fate. The dance revolves around the issue of the man's and woman's fate. She receives a rose that represents love, and he receives a noose that symbolizes death. In an ironical ending, she discovers that he is not her love, and he realizes that she was not his death. On a small set, there is a mock-up stage with curtains. Sallie Wilson enters from behind a curtain that resembles the temporary set-up of a fortune teller at a circus. She wears a blond wig, a scalloped chiffon cap, and a costume of gray, silver, black, and white, like a goddess-apparent. Her hand motions are robotic at times. She puts herself in between the couple, and occasionally shadows the woman (in a green dress), following her in *bourrées.* The couple embrace, then begin to fight. One of the hand-arm gestures is characterized by a bent elbow. The man (in a red jacket) seems to dismiss the woman while Sallie lures him to her den. He returns and focuses on the woman. They move back and forth quickly as if they cannot make up their minds. When Wilson forcefully reconquers the space, she has an aura of height and power. She helps lift the woman onto his back, walking on the diagonal, he with the burden on his back.

Tudor enjoyed depicting the danger of temptation and bad behavior. Several of his ballets revel in some kind of naughtiness or fall from good taste. In other words, evil and destructive qualities were understood by Tudor, who should probably have created a ballet role for the devilish Mephisto and then danced it. In the studio, everyone thought that was what he was doing anyway.

Les Mains Gauches became part of a larger piece, *Concerning Oracles,* when it was performed by the Metropolitan Opera Ballet fifteen years later in 1966.

One month after *Les Mains Gauches,* Tudor created another frothy piece, *Ronde du Printemps,* to an Eric Satie score based on the easy exchange of loving partners, a relay-style batch of seductions. During the same summer, Myra Kinch's dances, La Meri's exotic vignettes, and Ted Shawn's *Song of Songs* were shown.

Ronde du Printemps (August 1, 1951)

Walter Terry wrote: "Tudor's *Ronde du Printemps,* a series of *pas de deux,* is a larger, more ambitious work but not as successful theatrically as *Les Mains*

Gauches. Like Arthur Schnitzler's *Reigen,* a hearty series of seductions with one member of an amorous experience continuing his activities into the succeeding scene, provides the story line of this vernal round. A lady of easy virtue, a diplomat, a maid, a soldier and a society lady are among the several characters engaged in romantic dalliance. . . . But at the moment, it's too long and too thin. Occasionally Tudor uses the freshest, most pungent and engaging movements with subtle psychological inflections done in a light French comedy style."[14]

Sallie Wilson recalled that the choreography for *La Ronde* came out easily:

> It was so delicate, and full of wonderful, delicious laughs, I don't know why he didn't give it the stature that it should have had. We were just students, and he thought it was just a little thing. During rehearsals, Tudor wouldn't let us look at each other on stage, and before our entrance, he wouldn't let us see the sequence of the dancer who preceded us. This forced us to think seriously about our own characters and to listen more closely to the music.[15]

Unfortunately, *Ronde du Printemps* has never been revived.

In the fall of 1951, Tudor took on additional responsibility and began a prestigious teaching job; he became the head of the ballet division at the Juilliard School. Martha Hill, a dance visionary, was the director of dance and Martha Graham, whom Tudor felt a strong rapport with, was the head of the modern area. At Juilliard he devoted more time to teaching and caring for the development of dancers who had completed high school, but who wanted the intensity of a conservatory experience before dancing professionally. Years later, in 1963, Hill praised Tudor's dedication to the young students. "The ballets Tudor has 'arranged' for Juilliard have been an integral part of the development of the young dancers. In a performing art, apprentices must perform. His ballets have had freshness and wit, always with the Tudor hallmark of delightfully conceived and phrased movement and rightness of movement for members of the ensemble."[16] Martha Hill continues to extol the value of Tudor works and especially the quality of his teaching.

Despite a tight teaching schedule at the Metropolitan Opera Ballet School and Juilliard, Tudor managed to appear with Laing and Kaye when Ballet Theatre restaged Fokine's *Bluebeard.* He also was invited to restage New York City Ballet's *Lilac Garden* (November 30, 1951) for Nora Kaye, Hugh Laing, and Tanaquil LeClercq, a very talented young ballerina, soon to be Balanchine's wife. Kaye received excellent reviews as Caroline. John Martin wrote: "Kaye is altogether incomparable. A rich, warm womanliness pervades her characterization and subtleties of movement managed to convey her controlled desperation with utmost poignance."[17] Walter Terry echoed Martin: "The undeniable star was Miss Kaye. Every dance movement, every gesture revealed the tormented heart of the girl or mirrored the fleeting ecstasies experienced in the quick and furtive embraces with her beloved."[18]

That winter, Tudor created another full-length ballet for the New York City Ballet. Its premiere was in February, one year exactly after *Lady of the Camellias;* the music, a series of Beethoven overtures, created quite a stir.

La Gloire (February 26, 1952)

La Gloire received only a few New York performances, but later it was taken by the New York City Ballet on its European tour. It premiered in London on July 14, 1952. According to Cyril Beaumont, the ballet presumably derives its name from the play (1921) by Maurice Rostand and belongs to the same period.

The ballet was inspired by the film *All About Eve* as well as by memories of Tudor's bitter experiences in an ambitious and cruel dance world; little escaped his mocking wit, especially the kinds of feuds and vendettas associated with backstage and rehearsal behavior.

The set, conceived by Gaston Longchamps, satisfied the angling back and forth from stage to backstage with a device of shifting three semi-transparent wings painted in classic style. Pantomime and dramatic gesture characterized the movement vocabulary, and the corps participated like a Greek chorus, with predictions and warnings as they entered and exited.

La Gloire revolves around the decline of a great actress in a nineteenth-century European theater, with her remarkable moments on stage as Lucretia, Phaedra, and Hamlet at a time when it was not unusual for a great woman actress to portray important male roles. In addition, the ballet delved into her tense personal life and her flirtations with leading men in the wings and offered searing insights into the magical theater world.

But the star is haunted by the Dancer in Gray who will eventually succeed her. Just as the star returns for the final scene in her role as Hamlet, she shivers as she watches her understudy blatantly exhibit herself, gaining control over her roles in Racine's *Phaedra*, Arnault's *Lucrèce*, and Shakespeare's *Hamlet*. During the scene when Hamlet and Laertes duel, the Queen drinks from the poison cup, and the King is killed. With Hamlet's death, the symbolism of the star's demise emerges as the understudy steals from the wings in readiness to replace the star.

In *La Gloire,* "Nora Kaye . . . dominated the whole ballet with a

La Gloire, *1952. Hugh Laing and Nora Kaye in New York City Ballet's production.*

subtle appreciation of stage life before and behind the curtain."[19] To many, *La Gloire* was a mixture of brilliance and mistakes: "Marie Rambert told John Percival that it conveyed more of a backstage atmosphere than any other work she had ever seen."[20]

The music became a minor *cause célèbre* surrounding the ballet. Tudor courageously introduced Beethoven overtures for each of the scenes his great dramatic actress plays. Was Tudor in his right mind when he chose Beethoven's music for his score? Dance critics voiced dismay, and Tudor agreed. As Tudor explained to John Gruen in *The Private World of Ballet,* he usually preferred music that would be amplified and enriched by the choreography rather than hope that the strength of the music would save the choreography. He suggested that the music for his 1942 *Pillar of Fire* was not a completely satisfactory concert piece and that his choreography made a good marriage with the Schoenberg composition. In this case, however, Tudor confessed: "I made the terrible mistake of choosing three Beethoven overtures. When you get all the bombast of all those drums going full blast, then how do you match up to it onstage? You just can't go in competition with it. You can't really make music the partner. The music must dominate. So again, where are you without the music?"[21]

When Elizabeth Sawyer, Tudor's accompanist, asked, "Why did you choose Beethoven?" Tudor responded, "My dear, I just got carried away," and that was that. To cap off the story, Sawyer remembered an apt quote from Edwin Denby, who said that "Beethoven stamped on Tudor, and somebody had to pick up the pieces."[22]

John Martin was one of the critics who did not appreciate Tudor's choice of Beethoven's overtures, as he found them "hypnotically familiar, with independent connotations of their own and so full-bodied in volume that the visual action cannot stand up to it."[23]

But Sallie Wilson demurred. She spoke of *La Gloire* as a great ballet and bitterly complained that "John Martin, the dance critic of the time, panned the ballet and therefore it was doomed. *La Gloire* was incredible, most ingenious. First a scene on stage, then off stage. Nora Kaye would be having a fit in one part of the stage because her understudy was too energetic, and the audience could still see the action on stage. And while Diana Adams would be practicing some of Nora's steps over and over again, Kaye would be doing something and then suddenly and sinisterly notice her."[24]

Agreeing with Wilson, Doris Hering credited *La Gloire* for its "good sense" and both the tone and the theme of the "aging actress who must yield to the insistent grasp of time as represented by her predatory understudy. . . . Tudor fuses a matchless ability at dramatic gesture and dance impetus as, for example, the slow lifting of Phaedra's arm in a way that made the whole body grow in stature; and Hamlet's forward fall with the back turned to the audience and later on his arms stretching into space and seemingly pulling the body after them."[25]

Walter Terry grudgingly said that ten years after the opening of *Pillar,* *La Gloire* was not "all bad."

It is forced, empty and pompous at times and occasionally obscure in its development of an emotional sequence, but it has its moments of power and it has, basically, a stirring story line for its central figure. It is composed of episodes in the life of a great actress of the theatre. Bernhardt? Duse? It does not really matter—although it seems designed to parallel Bernhardt's career—but what does matter is that its narrative line is intended to carry the star from youthful roles through mature ones to character parts while mirroring the maturity, and the aging which accompanies this process. Tudor selects various points of glory that carefully focus the action. The actress is at her most magnificent at the close when she can no longer play young heroines and is given the role of Hamlet. Nora Kaye activates her role with gestures which speak of pride, arrogance, fear and dedication.

Terry found Nora Kaye commanding in her performance of this powerful personality, "where passages of bitterness and humor highlight her dancing."[26]

Like *Lady of the Camellias*, *La Gloire* was never restaged by New York City Ballet. Perhaps *La Gloire* contributed to Tudor's inglorious fate at New York City Ballet as, other than restaging *Dim Lustre,* he never again created a new work for them.

Five months after *La Gloire* opened, Tudor presented another ballet. If *La Gloire* revealed the backstage backstabbings of the theatrical world, *Trio Con Brio* was a sarcastic send-up of an egoistic ballerina.

Trio Con Brio (June 27, 1952)

Vispitin, the name of the choreographer listed in the Jacob's Pillow program for *Trio Con Brio,* was the *nom de plume* that Tudor chose in an informal bit of raillery. Tudor liked to tease and have fun with classical ballet, though as a young man with Rambert he took his technical training very seriously, especially as he started dancing late and recognized the difficulties in overcoming a late-blooming technique. But he also saw the classical vocabulary as an articulation of aristocratic snobbery and a perpetuation of mannerisms that deserved deflating. His most cutting exploration of these stances occurred in his ballet *Gala Performance,* but he gave it another go in *Trio Con Brio.*

In this short, punchy *pas de trois* technical statement, Tudor presents a woman (originally Tatiana Grantzeva), in a short white tutu with the requisite crown, the archetype of the Russian ballerina, and two men (Nicholas Polajenko and Ralph McWilliams) in short velvet jackets and white tights.

They work in Petipa-like movement formulas to Glinka's predictable music with pirouettes, *bourrées,* and movement patterns. The men partner her and walk around like peacocks. She acknowledges their obeisance, treating both men as elegant porters for her leanings. Upstage, she extends her leg front and arches back. She waits and deigns to move from one cavalier to the other, balancing, pirouetting, and unfolding her legs. The men have their moments of individual triumph as well. In a very presentational way, they cover the jumps, turns, and the quick, difficult *batterie* that we associate with the origins of ballet. Alone, the ballerina executes step *piqué* pirouettes and *pas de chats* on diagonals. She concludes with a double pirouette and bows as if to say, Now the audience must recognize my command performance. The show literally stops, and she goes off. After the other cavalier performs his variation with a lot of quick, stiff, beating jumps, the three dancers come out with typical end of the *pas de deux/trois* fireworks—*ballonnés, czardas* steps and rapid folksy steps, one arm up, pulling into fifth positions, a fitting ending to this imperial academic display.

Trio Con Brio was later performed in Hartford, Connecticut, on December 7, 1952, with other dances by La Meri and Ted Shawn doing the *Mevlevi Dervish.* The program toured Washington, D.C.; Pittsfield, Massachusetts; Rye, New York; and Vineland, New Jersey. A newspaper article mentioned that Grantzeva suffered an ankle sprain in Vineland. Though one critic spoke of considerable audience applause for Tudor's piece, in the Pittsfield newspaper (October 1952), the writer added that "Tudor's *Trio con Brio* was controversial. It was amazingly difficult to dance, it seemed very long and pretentious, but revisions have been made on it and the result is a happy one indeed."

Trio con Brio has not been restaged since 1952. A Jacob's Pillow film exists in the Dance Collection in New York City.

In the spring of 1953, on March second, a BBC British television program, at the time the sole English television channel, presented Tudor on "Ballet for Beginners." The program was initiated as a series scripted by Felicity Gray designed to promote understanding and appreciation of ballet. According to Janet Rowson Davis, the whole of *Jardin aux Lilas* was given, along with substantial extracts from *Soirée Musicale,* as well as Part Two from *Dark Elegies.*[27]

The winter of 1952, Tudor busied himself running the Metropolitan Opera Ballet Company and teaching at both the Met's Ballet School and Juilliard. Tudor organized lecture demonstrations for the final project in some of his courses, creating pieces that reflected the students' talents. Some of these showcases were more lavishly produced than others. Tudor's

Exercise Piece, which lasted nearly twenty minutes, used nineteenth-century composer Ariaga's String Quartet No. 2 in A Major. It was preceded by Tudor's lecture, "Let's Be Basic." *Two by Tudor,* presented May 7 and 8, 1953, at Juilliard, was essentially a lecture-demonstration in which Tudor commented on and described various elements of the ballet technique.

Exercise Piece (May 7, 1953)

Tudor shared this event at Juilliard with Doris Humphrey, who organized the second half of the program, entitled "What Dances Are Made Of." Humphrey, a distinguished modern dance choreographer, taught dance and choreography at Juilliard. On the occasion of *Two by Tudor,* Tudor complimented his outstanding students, those whose work he particularly admired, such as Carolyn Brown, who became Merce Cunningham's lead dancer and was a frequent figure in Tudor's ballet classes at the Metropolitan Opera Ballet School.

Caroline Bristol Britting, one of the original dancers in the ballet, described the costumes: "The men were dressed entirely in gray, with tights, short sleeved shirts and ballet shoes. Some of the women wore pale-yellow tunics; they were on pointe. Others were in mustard shades and the largest group of women wore gray tunics and ballet slippers." Bristol remembered that Tudor arranged separate groups of six, four, and so on, to perform classroom ballet combinations. The sextets executed various adagio movements and partnered one another. There was also a quartet of four short women; one woman danced two measures behind the other three trying to keep up with them. In the course of these displays, Paul Taylor entered and exited, shifting here and there, doing his own quirky, modern movements, totally outside the ballet realm. Taylor looked immense in comparison to the quartet of diminutive ladies, providing comic relief. The larger groups represented the corps de ballet and wove themselves into intricate patterns that changed numbers and directions. During the finale, all twenty-six dancers came together in a grand ensemble finish.[28]

"*Exercise Piece* was what its name implied, an exhibition of technical combinations used for the training of dancers. As such, it was neatly and cleverly crafted."[29] Parts of *Exercise Piece* were notated by Juilliard students; the notation can be found at the Dance Notation Bureau.

That summer (1953), Tudor returned as usual to Jacob's Pillow for his summer teaching and choreography. He composed a charming dance that evoked the wistful innocence and reverie of children's games, setting it to Robert Schumann's equally nostalgic score, *Kinderscenen.*

Little Improvisations (August 28, 1953)

If we imagine the games of children *cum* adolescents on a rainy day, their gestures, responses, and dress-up games straddle a fine line between sexual awakenings and innocent child's play. Tudor indicated that he wanted two dancers who looked quite young, perhaps ten or twelve years old. He sought a casual distinction, a subtle and amusing interaction of movement and gesture for Schumann's evocative score.

The ballet begins with two people sitting on a bench upstage center. Arms opening out, yawning, the girl, Yvonne Chouteau in the original production, wears a little skirt, a ponytail, and a bow in her hair. The young man, originally Gilbert Reed, in a white shirt and white pants, partners her while she stands on the bench lifting her leg high and to the side. They do Hungarian folk steps, and he slides her to the floor. He then moves up stage to sit on the bench with his back to the audience while she executes a series of charming movements that quickly change direction and take her rapidly through space. She removes a cloth from the bench and covers her head as if it were a scarf. They begin to dance with it, and coyly she wraps it around him like a tunic. He leaps into a typical male variation, legs beating and big jumps and pirouettes, into a "my heart bleeds for you" fall to the ground. She grabs the cloth and folds it into the shape of a baby, cradling it. She gently sways with the "baby," her head tilted in its direction. He holds the baby, too. They next play at being *danseurs nobles*. While he supports her, she lifts one leg behind her and dips forward. She opens the cloth and fashions a little animal. When she wears the cloth as a cape, he leads his queen to the bench where she sits and commands him to kneel before her. After they dance together, he replaces the cloth and sits down on the bench. The dance finishes when she kneels in front of him and rests her head on his knee.

Walter Terry spoke of the ballet as "a completely captivating work enlivened by the improvisational air suggested by the title and its basic sweetness and appealing innocence spiced with dashes of humor. Although it is simple, it discloses the Tudor inventiveness in every subtle phrase of action. Whether its beauties would be lost in a large opera house, I could not say. Here was Great Tudor!"[30]

Wayne Smith commented that *"Little Improvisations* gave fragmentary glimpses into the lives of children and showed their un-selfconscious play together as a soldier and a little mother, or a princess and a Knight. The work had all the simplicity of great art!"[31]

Nils Åke-Haggböm, director of the Royal Swedish Ballet, spoke animatedly about dancing in *Little Improvisations:*

Little Improvisations, *1962. Sirpa Jorasma and Eric Hampton from Juilliard.* (Photo by Elizabeth Sawyer)

> [Tudor] was very impatient with me. The rehearsals were intensely concentrated. In Sweden the ballet was called *It's Raining* as the weather in Northern Climes is insufferable. When seated with my back to the audience, while Annette Av Paul, my partner, was doing her solo, I thought I could take it easy and rest. Not at all! Tudor tore apart my seated position. How is the chest? Where is your head? Nowhere? The back and neck are your face. Don't go to sleep! I wasn't sure I would survive these rehearsals.[32]

In another review of the ballet, Wayne Smith extolled the piece. "Tudor's new duet was . . . an exquisitely lovely emotionally tender work created with the supreme simplicity of great art."[33]

Little Improvisations was restaged by the Norwegian Ballet (1961), Juilliard (1962), the Royal Swedish Ballet (1963), and many smaller companies, including the Tokyo Civic Ballet in 1990.

That fall, on November 18, 1953, Tudor re-created *Gala Performance* for the National Ballet of Canada in Ottawa with remarkable success.

In December, Juilliard presented a special week-long program: "The British Festival was a major effort that involved the entire school in preparation and performance of a week-long series of concerts and theatrical productions. There was a moratorium on all regularly scheduled classes to permit rehearsals and public performances. A true Festival spirit permeated the school."[34]

Elizabethan Dances (December 7, 1953)
Britannia Triumphans (December 11, 1953)

The suite of Elizabethan dances was performed in collaboration with Suzanne Bloch and her group of Early Music students. Costumes were designed for the musicians and dancers, authentic musical instruments were played, and the Tudor rose decorated the dancers' costumes. "Here, with limited pupil resources, Tudor had to reconstruct dances of the Elizabethan period, evoking the atmosphere of a theatrical genre long since vanished." The dance critic Doris Hering indicated that this work proved Tudor "at home in any century and with any style of characterization."[35]

The British festival proved a resounding success: "Tudor's creative accomplishment was monumental; young students not yet fully trained; a very complex production requiring many facets of collaboration with musical forces, dramatic elements and staging problems which used the house and its aisles as well as the stage."[36] Though the dances have never been reproduced, excerpts from the suite have been notated and may be found at the Dance Notation Bureau in New York.

Just before the performance of *Britannia Triumphans,* Tudor had to quickly return to England, as his mother was extremely ill (she subsequently died). In a letter to Tudor in England from Martha Hill dated December 21, 1953, she wrote that Lincoln Kirstein and Edwin Denby attended the concert at which *Elizabethan Dances* was performed and they were very excited about them. Hill stated that "they thought it would be nice to do them for the New York City Ballet." Nothing seems to have resulted from this casual remark.

While Tudor spent most of his time in New York City, occasionally he accepted jobs elsewhere, and in 1954 he agreed to create a new piece as a one-night stand with the Philadelphia Ballet, where he had been teaching and getting to know the dancers. He choreographed *Offenbach in the Underworld* to the same score as Massine's *Gaîté Parisienne* (1938), a well-known piece of music that represented the frivolous, frou-frou, music-hall works Tudor often deprecated.

Offenbach in the Underworld; or, Le Bar du Can-Can (May 8, 1954)

The Offenbach ballet was made at the behest of the Philadelphia Orchestra in 1954 for a semi-amateur dance group in Philadelphia. Since the ballet proved a rousing success, Tudor remounted it for Ballet Theatre in 1956. It was performed later that year on their tour to Covent Garden with the same choreography but with a lesser-known Offenbach score. An indignant Massine forced Tudor to use different music, as he was angry that Tudor had the audacity to choose the score for Massine's world-famous *Gaîté Parisienne* originally choreographed in 1938.

The atmosphere in *Offenbach in the Underworld* suggests a colorful Parisian cabaret full of odd types and ill-assorted lovers. Perhaps it is this ballet that Alexander Bland was thinking of when he wrote the "Tudor was always the most French of English choreographers."[37]

The Ballet Theatre program, from April 18, 1956, stated:

> The action of the ballet takes place in a fashionable café in the eighteen-seventies, where many people come to relax and enjoy themselves at the hour when most people go to sleep. As the café has an international reputation, visiting celebrities come to see, to be seen, and to be amused. Among these are a famous operetta star and one of her admirers, a grand duke. A debutante of good family, accompanied by three of her friends, arrives veiled in order not to be recognized in a place which for families is "Out of Bounds." There is a painter who, like all young painters, is penniless and tries to earn a meager living by drawing sketches of the café's patrons. There is no story to the ballet, for the flirtations that take place at such a place at such a time, are most often half-forgotten by the next morning. There is neither a sad ending nor a happy ending, but only a closing time.

The action during the ballet demands that the protagonists display their characters with a great deal of panache and sensuality. If one couple happens to glide into another couple, that physical connection arouses sensual sentiments beyond one's partner. Deborah Jowitt commented that "Tudor probes at the stereotypes with vicious cynicism."[38] The comic opera diva, for example, tried to seduce whoever is game, especially the Grand Duke. Jowitt reminded us that "their dance together is a travesty of the rapturous waltz of the Baron and Glove-seller in *Gaîté Parisienne.*" The Duke's behavior as a representative of aristocratic meanderings or fooling around in Paris descends to painful comedy. The women who cajole the Duke are resurrected from the tough and determined ladies of his *Judgment of Paris* or from the decrepit street ladies in *Undertow*. The romantic young painter engages the glittery Operetta star in a swirling moment of passion. A rousing can-can

Offenbach in the Underworld, *1956. Ballet Theatre's production.* (Photo by Sedge le Blang)

energizes the stage but without the absurd, ferocious joy of Paris after the Commune. The conclusion of the ballet brings a young girl, an innocent child (daughter of the Patronne) in a nightgown into this decadent café. People lying on the floor, prostrate from their excesses, gradually are touched by the child, arise from their satiety and depart.

Responses to the ballet were mixed. When originally performed in Philadelphia and later in Brooklyn, *Offenbach* was found to be a "work of wit, charm and atmosphere." But John Martin alleged that when it was restaged for Ballet Theatre, there were not enough rehearsals; thus, "it lacked spirit and the ensemble lacked tightness. For comedy that is all but fatal. The can-can which is potentially the rowdiest and wittiest thing of its kind on record, and the knock-down-and-drag-out fight were better done by the Canadians. With time, and repeated performing, it will emerge as the charming work it really is." Martin, who paid no attention to the little works Tudor created at the Pillow and Juilliard, reminded his readers that this was Tudor's first new ballet in four years.[39]

In a brief *New York Times* clipping on the twenty-sixth of February, 1955, an event was advertised in which Tudor's "can-can" from *Offenbach in the Underworld* was performed. The Philadelphia Art Museum sponsored an arts festival with the Police and Firemen's Band honoring Grace Kelly and Marian Anderson. There were 15,000 people in the audience.

When *Offenbach in the Underworld* was performed by the Philadelphia Ballet in 1954, the results were so encouraging that Tudor and Nora Kaye took the ballet that summer to Tokyo, where it was performed in a revised version by the Komaki Ballet with Miss Kaye as the Operetta Star. On August 25, 1954, in the Nichigeki Theatre, the opening of their season included *Jardin aux Lilas* and *Offenbach in the Underworld*.

Subsequently, Tudor mounted it for the National Ballet of Canada on January 17, 1955, at the Palace Theatre, Ontario. The Canadians repeated this performance at the Brooklyn Academy of Music on March 26, 1955. Celia Franca, who was in the original cast of *Dark Elegies* in London, played the Operetta Star in Canada. The Joffrey Ballet played it in 1975 and 1983.

Tudor continued to travel around the world re-creating his ballets, rather like a prophet promoting his gospel. During his visit to Japan, Tudor was impressed by the beauty of its ancient shrines and the philosophy of Zen Buddhism. It was an enchantment that perhaps began seriously during this period and continued for the rest of his life.

Tudor's relationship to Ballet Theatre remained tenuous, although they continued to stage and perform his ballets. Happily, Lucia Chase and Oliver Smith were still committed to having full evenings of his work each season. John Martin wrote that "enthusiastic cheering for all" greeted the performance of Tudor's *Judgment of Paris, Romeo and Juliet*, and *Pillar of Fire*.[40]

During the summer of 1955, Tudor was invited to the Athens Festival to help stage dances, although he never mentioned just exactly what he did in his rapturous letters home, except to complain about rehearsing untrained dancers. His letters sang the pleasures of being in Greece and of sharing its ancient glory. In the autumn, Tudor returned to his teaching at Juilliard and the Met but took time out to lecture in October at the University of Toronto. The previous March, the Ontario National Ballet performed some of his works, including *Offenbach in the Underworld,* at the Brooklyn Academy of Music.

Pas de Trois (April 25, 1956)

Tudor created *Pas de Trois,* which was performed first on April 25, 1956, by Juilliard students Caroline Bristol, Gail Valentine, and Bruce Marks at a

Snapshot of Antony Tudor and Nora Kaye, ca. 1954, in a playful moment. (From the Performing Arts Research Library)

Festival of the Arts held at the International House in New York City. The trio then danced it again for their graduation requirement on May 5 at the Juilliard School. It was restaged at Juilliard the following September, with Kevin Carlisle dancing Bruce Marks' role.

Caroline Bristol Britting recalled that the men were dressed in blue tights and shirts, while the women wore dusty-rose long-sleeved leotards and bright pink tutus with pink flowers in their hair. The dance began with a series of quick allegro combinations, followed by several *pas de deux* where Marks partnered each of the women separately, and then together. Everyone danced a short solo tailored to his or her talents and individual capabilities. Marks flew around the stage doing large jumps and turns, Bristol did small, rapid beats and turns, while Valentine was given more legato and lyrical movements. For the finale, all three danced together to finish.[41] The piece was partially notated by Bristol, and the score may be found at the Dance Notation Bureau.

Another big Tudor evening, a momentous event at Ballet Theatre, occurred May 1, 1956, when the company celebrated Tudor's twenty-five years as a choreographer. *Undertow, Romeo and Juliet,* and *Offenbach in the Underworld* received dedicated performances. Just one week before the performance, John Martin wrote about Tudor's importance as an historical dance figure. His *New York Times* column on April 22, 1956, once again traces Tudor's heritage to the genius of Fokine and at the same time extols the subtle changes and unique contributions Tudor made to the ballet stage:

> In a very real sense Tudor carries on the broad tradition of realism that was the credo of Fokine's great revolution against artificiality and abstraction. But he departs in almost every respect from the romanticism, the ethnological interest, the tendency toward dramatic generalization typical of Fokine. His dramatic direction is inward toward the specific and the psychological, toward believable tensions instead of generic situations. It is like modern dance but it avoids the subjectivity that actuates much of the modern dance and the personal movement [is] highly original and colored at every turn by character situation, creator's intention . . . his canny way with music, devastating sense of satire, his use of period styles, his slow and careful craftsmanship and the wide range of his subjects.

Martin recognized Tudor's refined sensitivity and suggested that only George Balanchine's work could be compared in value to Tudor's.

There was an active effort in the next few years to stage Tudor's ballets in Scandinavia. On November 15, 1957, the Royal Swedish Ballet put on *Gala Performance,* with Peggy van Praagh directing the restaging. In 1957, the Norwegian Ballet at the Folketeatret in Oslo danced *Soirée Musicale,* and in 1958, Peggy van Praagh reworked *Judgment of Paris* in Oslo. Later, in 1961, Tudor would stage *Jardin aux Lilas* for the Norwegians, and *Pillar of Fire* and *Dark Elegies* for the Swedish Ballet.

In between these travels, Tudor continued to work with the Juilliard and Met dancers. He also began his association in 1957 with the First Zen Institute in New York City.

Mary Farkas, who helped found the First Zen Institute, recollected in the English journal *Choreography and Dance* that Tudor often wandered into prayer meetings at the Institute in the late 1950s.

> In the beginning he would come to the meetings in Greenwich Village, often late as he was very busy with his dance world. I particularly remember his coming after meditation had already started. When you would go a certain distance into the room, you could plainly feel the silence. He would freeze at that point and stand in meditation immobile as a statue until the period was concluded.

She described Tudor when he grew into an important figure at the Zen Institute, which later moved uptown to Thirtieth Street. "In the 60s and

70s, bolt upright, thin-lipped, bald, robed in black, Tudor could easily pass for the head of a monastery. . . . Timing and discipline built in, his movements commanded respect, even awe."[42]

Tudor was aging. He had no family, and the central relationship in his life, his friendship with Hugh Laing, continued, but they were no longer as close as they had been. When Tudor became a Zen follower, he discovered a way to quietly and purposefully meditate and ponder the truths of our inward experience as well as everyday life. Furthermore, Zen may have been an escape from the aridity of his professional and private life at this time.

Ballet Theatre suffered a serious loss in July 1958 when a trailer truck carrying Ballet Theatre properties caught fire near Nice, France. Sets, costumes, and scores for sixteen ballets, including some of Tudor's, were destroyed, the loss being estimated at $400,000. Though they received emergency funds from the State Department and sets and costumes from European ballet companies, the damages were costly and difficult to repair. The expense of reconstructing the scenery of old Tudor ballets may well have influenced future decisions about redoing his works.

A record-breaking trip to South America, where Ballet Theatre toured for more than three months, brought Tudor to Buenos Aires at the Teatro Colon. There he staged, on August 19, 1958, a ballet, *La Leyenda de Jose* (*The Legend of Joseph*), to the music of Richard Strauss. Nothing more is known about it. Oddly enough, in a record of the Teatro Colon's performances at the New York Public Library, this ballet was not mentioned. Tudor also re-created *Pillar of Fire* on the same date, and there *is* verification for this.

After several years of movement silence, 1959 was a breakthrough year for Tudor, for, if he did not make another major ballet, he was at least choreographing again. He collaborated with filmmaker Maya Deren on a creative work, *The Very Eye of Night,* as his interest in the camera and filmmaking extended from his early years experimenting with the BBC. The fifteen-minute film had its premiere at the Living Theatre on February 9, 1959, and continues to be a popular avant-garde film. Maya Deren oversaw the many complex elements that go into a movie, such as the music, by Teiji Ito; the lighting, directed by Ernst Neukanen; the sound, recorded by Louis and Bebe Barron; various camera workers; and securing the cooperation of the Metropolitan Opera Ballet School.

Tudor enjoyed contemplating how planetary relationships affect human experience. His ballet *The Planets* (1934) also explored mortals under the influence of heavenly bodies. The radiances of one sort of planet would tug and pull, having a powerful impact on certain kinds of personalities.

Program notes for the film include the following:

Maya Deren, Filmmaker, ca. 1959.

We fall asleep; the laws of micro- and of macro-cosm are alike; travel in the interior is as a voyage in outer-space; we must in each case burst past the tension of our surface—our here-space and our now-time—to enter worlds measured by light and sound. This film lives in the world of this idea; this idea can live only in the world of film. *The Very Eye of Night* is a Ciné ballet of night filmed in the negative, creating the illusion of movement in unlimited space, the dancers resembling sleep walkers become four dimensional, advancing as if planets in the night sky. . . . This is a metaphysical, celestial ballet of night. The blackness of night, as the opposite or apposite of day, erases the horizontal plane of the earth's surface. By day we move according to desire and decision; by night Noctambulo is moved by gravities. Advancing with the blind, incalculable accuracies of a sleep-walker, led by the twins Gemini (as eyes are twinned, or as I is twinned), he is drawn to the celestial center which revolves eternally in the dark geometry of its orbit from the beginning of time. . . . Uranus (father of heaven) is the seventh major planet; Urania, his mythological female counter-part, is an epithet for Aphrodite; the names of the four satellites as known in astronomy, have been adopted for the four cardinal points in this film; and man, creature half of day and half of night, half of heaven and half of the abyss, is both contained in and contains this totality.

Since the brief glory of the London Ballet in 1938, Tudor had longed to direct his own company. Though that dream was still out of his reach, in 1959 he saw the time as ripe for the initiation of full ballet evenings at the Metropolitan Opera. As idealistic as this enterprise seemed, he was able to garner enough support from Rudolph Bing, general manager, and John Gutman, assistant manager, for an opening in March 1959. He invited John Butler, Herbert Ross, and Alexandra Danilova to create pieces for the concert, and fortunately he was able to persuade Nora Kaye and Lupe Serrano to dance in his new piece to Richard Strauss's *Four Last Songs*.

Hail and Farewell (March 22, 1959)

Was Tudor *Hailing* the birth of a new Met Ballet Company, or was he saying *Farewell* to Nora Kaye, as she had just married Herbert Ross, already a notable choreographer? They were working on the foundation of a new ballet company, the Ballet of Two Worlds, which planned to go on tour to Europe in the fall of 1960.

The Met occasion was historic as Tudor put together an entire program of dances expressly for the Metropolitan Opera's own ballet contingent. The desirability of such an event was long evident and its realization cause for rejoicing. After all, European opera houses poured large sums of money into their ballet companies. Why not here in America? The audience was filled to capacity to celebrate this exciting evening. The innovative choreographer John Butler created *In the Beginning* to Samuel Barber's First Symphony, and Herbert Ross made *The Exchange* to Francis Poulenc's organ concerto; Alexandra Danilova, *Les Diamants;* and Tudor, *Hail and Farewell.*

Unfortunately neither Butler nor Ross received very good notices. John Martin testily suggested that "both of these ballets might have been symphonic ballets that Massine forgot to create." The climate changed with *Les Diamants,* choreographed by Alexandra Danilova to the music of Charles de Beriot. Danilova's piece, with the bravura dancing of Lupe Serrano, took on added charm and chic from the elegant costumes of Karinska. Martin noted that Nora Kaye, in a moving solo passage in *Hail and Farewell,* "lifted [Tudor's] choreographic conception to genuine heights."

Martin pointed out that the purpose of the evening, which was to develop the talents of the Met Ballet, was hardly fulfilled: "The members of the Opera Ballet itself had little opportunity except for Bruce Marks, who danced in Butler's and Danilova's pieces. Edith Jerell handled a solo in Mr. Tudor's reflective setting of the *Four Last Songs* of Strauss with style and eloquence."[43]

A more enthusiastic Walter Terry praised the evening by saying that Tudor's title, *Hail and Farewell,* was a bit lighthearted for the highly poetic evening: "It turned out to be the best Tudor choreography in a good many years. . . . Though the opening Serenade seemed like a dancing school recital, the songs offered four gentle but gloriously styled solos. Unforgettable was Miss Kaye's dancing of the last song, a triumph she shared with the opera singer, Eleanor Steber. Mr. Tudor created glowing, choreographic images, haunting and exquisitely lyrical and they were brought to superb dance realization by all four soloists."[44]

In *Dance Perspectives* Selma Jeanne Cohen reported that the ballet evening, and notably Tudor's new work, *Hail and Farewell,* brought Tudor only partial success:

> After the preliminary academic section, the four solos received mixed reviews. Some audience members were disturbed by the lack of dramatic intensity, and climax qualities they had come to associate with great Tudor choreography. Here the treatment was lyrical, depersonalized almost to the point of abstraction. The movements were lovely, the phrasing musically beautiful; but the emotion of the Strauss songs seemed dissipated. Since the other ballets on the program were unsuccessful, the critics put Ballet Night in the class of ventures with good intentions gone astray.[45]

For Tudor, the appearance of the Rambert ballet at Jacob's Pillow during the summer of 1959 must have been as unsettling as a haunting from the past. Under the supervision of Tudor, the Rambert Company re-created *Gala Performance* and *Dark Elegies*. Several films were made of these performances. The Rambert Ballet received splendid reviews for their Tudor ballets, especially his *Gala Performance*. Strangely enough, in Rambert's biography, *Quicksilver,* she makes no mention of meeting Tudor on her maiden trip to the United States. Instead, Rambert enthusiastically recalled:

> In 1959 we were invited to appear at Jacob's Pillow near Boston. . . . Jacob's Pillow is run by Ted Shawn for long summer dance Festivals, to which he invites interesting dance groups of most varied types—classical, contemporary, short ballets, concert numbers and so forth. During the day it is a kind of University of the Dance, with a choice of brilliant teachers, and in the evening there are performances in their delightful barn theatre. It was lovely to have Martha Graham and Agnes de Mille come and see us and renew our friendship.[46]

—Quite a contrast to the bitter days in 1937 when Rambert dismissed de Mille as a meddling, non-classical American dancer.

How wonderful it would be to know if Rambert and Tudor spent time with each other at the Pillow discussing their fervent years together in the 1930s. Whatever they spoke of, it seems that she never quite forgave Tudor for leaving her and establishing a competitive company.

As we have seen, Tudor's customary relationship with Ballet Theatre usually involved restagings of his ballets and rehearsals for those productions. Yet Joel Kasow, in *Dance Chronicle*'s updating of the *Dance Perspective*'s Tudor Chronology, mentions the fact that during the 1959–1960 season at Ballet Theatre, "most of the living choreographers represented in the repertory came in to rehearse their own works, with the exception of Tudor, so that only Tudor's suffered in performance."[47] During this problematic season, the critics feared that the company, once "vital and brilliant," had burnt itself out and "virtually collapsed before our eyes." Major criticism centered upon Lucia Chase's ability to lead the company out of its doldrums.

Not without a certain sardonic humor, Tudor put together a group of Metropolitan Opera ballet dancers to make fun of opera ballets and especially opera themes for the occasion of a Metropolitan Opera Guild meeting on October 14, 1959. Dancers in the brief presentation recalled that evening as a hilarious romp. Tudor recreated *Ride and Fall of the Valkyries* to the Wagner score, his dancers dressed in Viking attire and lances, flinging themselves across the stage and into the wings. They sang as well, or rather shrieked. Here was the other side of Tudor, the witty, biting satirist whose comedic talents depended on a clever and highly developed, sarcastic sense of humor.

Tudor's long association with the Metropolitan Opera Ballet Company was always a compromise. Naturally the opera came first for the management, and the ballet far behind, and the Met certainly was not concerned with the old ballets Tudor had created for Ballet Theatre or New York City Ballet. Therefore Tudor alone was responsible for the perpetuation of his important ballets. Tudor persevered in his attempts to hold on to his repertoire, restaging *Little Improvisations* (1953) for Juilliard dancers in April 1960. At the same time, he continued his personal dialogue about the subject of ballet and the scope and range of its vocabulary, offering his Juilliard ballet students an opportunity to make this exploration with him. He collected an interesting group of pieces about the ballet technique, really very funny in concept, aptly titled *A Choreographer Comments,* to Schubert's Octet in F Major.

A Choreographer Comments (April 8, 1960)

Comment I: Arabesque—A position in which the body is supported on one leg, while the other is extended in back with the arms harmoniously disposed.
587 arabesques

Comment II: Jeté—A spring from one foot to the other.
224 jetés

Comment III: Pas de Bourrée—Three transfers of weight from one foot to the other.

Comment IV: Tour—a turn.
60 turns

Comment V: Quatrième en l'air—Leg extended in front.

Comment VI: Bourrée Courue—Small running steps.

Comment VII: Petite Batterie—Small jumping steps in which the legs beat together.
597 beats

Comment VIII: Posé—A step onto a straight leg.
65 posés

Comment IX: Tour—A turn.
184 turns

Comment X: Pas de chat—Literally, step of a cat.
1 pas de chat

Just typing out the program makes one giggle. Imagine counting 597 beats; but of course Tudor must have enjoyed that task. Selma Jeanne Cohen testified:

> Of the later Juilliard pieces, *A Choreographer Comments* probably had some of the wittiest passages Tudor has devised, notably the snidely priggish exposition of the pas de bourrée and the amusing spoof on *Swan Lake* labeled "1 pas de chat." Much of the rest was technically ingenious, though some of the audience—after seeing 587 arabesques followed by 597 beats—felt willing to accept Tudor's assertion that ballet had variety without watching the evidence slowly unroll.[48]

Tudor worked very well with certain of his students at Juilliard. Among the more favored was Pina Bausch, whose work interested Tudor from her early Juilliard days. It was during his rehearsals for *A Choreographer Comments* "that . . . I turned Pina Bausch into a comedienne."[49] Soon after, he arranged for Bausch to dance with the Metropolitan Opera Ballet Company. She subsequently returned to Essen and then Wuppertal in Germany, where she directed her own dance company and became one of the major European choreographers in the latter part of this century.

A Choreographer Comments begins with three women in short white skirts, joined by three more women and two men who support them in arabesque balances and lifts. Imaginative combinations of arabesques and lyrical adagio movements seamlessly unfold with a couple embracing at the end. Some of the *jeté enchaînements* remind one of those at the entrance of *Dim Lustre.* During the *pas de bourrées,* Bausch wears heels and a high headdress, while her partner, Koert Stuyf, sports a tricorne hat. The turns pass quickly on the diagonal; one woman, then another follows. When the group performs the *quatrième en l'air* exposition, the women wear short skirts, and they goose-step here and there, lifting one leg in front of the other in solidly vertical positions. Running steps for the *bourrée courue* have the young woman skittering all over the studio. The jumping *batterie* contains dozens of different *sautés, changements, brisés, cabrioles,* etc. The dancer, who poses with what seem like hundreds of steps onto pointe, finishes by limping off stage. The hamming it up continues as four women execute 184 turns; the *piqué* turns are especially funny. For the one *pas de chat,* Tudor puts together a charming composite of the vocabulary, recreating some *Swan Lake* tipping heads and rubbing shoulders. A most ingenious display! Excerpts from *A*

189

Choreographer Comments were performed at Juilliard in 1963, 1964, 1983, and 1984.

In 1960, Tudor created an intriguing and valuable documentary film, *Modern Ballet,* with his dancers Nora Kaye and Hugh Laing. Today the film serves as a useful instructive device in order to understand Tudor's working process. In the film, both Nora Kaye and High Laing sensitively express their characters by demonstrating moments in *Jardin aux Lilas, Gala Performance, Pillar of Fire, Romeo and Juliet, Dim Lustre,* and *Undertow.* Their performances from *Romeo and Juliet* are particularly moving. In *Pillar of Fire,* Tudor dances the role of the Suitor with Nora Kaye. One notes the power and vigor of Tudor's dancing. He was already a balding fifty-two-year-old man, but he was able to lift Nora Kaye firmly and to impart a strong sense of stability and tender emotion. He gives a moving performance in those short clips. The film was originally part of a television series, *Time to Dance,* initiated by Martha Myers, who interviewed various choreographers about their dances, some of which would also be performed on the show. Questioned by Myers about his choreographic methods, Tudor speaks eloquently about his manner of working, the subtle way he would use a walk to denote a dancer's characterization, or the manner of phrasing groups of movements to portray an action.

In the fall of 1960, Tudor traveled to Munich to the Prinzregenten Theatre, where he helped restage *Gala Performance.* He revisited Stockholm, and in Oslo he re-created several works, including *Soirée Musicale.*

Surprisingly, bravely, that winter Tudor tried his hand at choreographing for the Metropolitan Opera once more. He made dances for two operas, Gluck's *Alcestis* (premiered December 6, 1960) and Wagner's *Tannhäuser* (premiered December 17, 1960). "His dances were unsuccessful, though John Martin urged that both contained some first rate choreography and that Tudor should have been blamed not for being dull, but for allowing himself to appear at such disadvantage in a production so unsuited to his talents."[50] Corps members remembered wrestling with the Venusburg scene in *Tannhäuser,* where there was so much commotion in deep darkness that at one point one of the dancers was being lifted and carried off by the wrong partner. They also described some wonderful movements in *Tannhäuser* where Tudor created surging and tangled groups fusing with one another and then dissipating.

Arthur Todd in *Dance Observer* described Tudor's inspired choreography for *Alcestis.* "There are many instances where dance movement and singing are so well-integrated that it is difficult to discern the difference between the singers and the dancers."[51] Several months later, Todd wrote "Ballet at the Met" in *Dance Observer,* in which he praised Tudor's *Tann-*

Alcestis, *1960 production with Nicolai Gedda and Eileen Farell. From left to right: Suzanne Ames, Judith Chazin, Patricia Hayes, Jeremy Blanton, and Wally Adams.* (Photo by Louis Melancon)

häuser for its "voluptuous and sensuous choreography." He described the dances as being "far from classical and definitely not incidental," such as one usually finds in opera.[52]

Tudor echoed Todd's remarks about the role of the choreographer in the integration of opera dancers and singers in "Movement in Opera":

> My own aims are best satisfied when the production achieves such a sense of theater that there is no discernible division of labor or effect between the respective contributions of director, designer and choreographer—and we must include the performers, since finally it is they who have to make the synthesis live.[53]

Despite Tudor's purposes, we know that true collaborations of that magnitude rarely occur, especially in such a hidebound and hierarchical organization as an opera company.

Undeterred by the mixed reviews at the Met, Tudor displayed a new creative vigor, producing an interesting ballet for his dance students at Juilliard that spring.

Dance Studies (Less Orthodox) (May 8, 1961)

Dance Studies (Less Orthodox) was re-presented in a larger work, *Gradus ad Parnassum* on March 8, 1962, with several other choreographers from Juilliard. According to Selma Jeanne Cohen, audience members remarked that this piece was less direct in its humor than *A Choreographer Comments,* and rather ambiguous in its intent. "But musicians marveled at Tudor's sensitive and subtle use of the complex music of Elliot Carter."[54]

At the opening of this piece, Tudor catches the audience off guard. The stage is designed with an unlikely set that reminds one of a circus tent, with swinging ropes and a ballet barre upstage. Two men and ten women dance in this piece. The men are dressed in black shirts and tights and white ballet shoes, while the women wear short tunics. Then, unexpectedly, a big gorilla enters for a short time while one woman performs an adagio combination. Two other women move to the barre where they form sculptural poses and some quick moves. As one woman dances, the others place her on the barre and carry her off stage. In another scene, four girls in leotards and short draped skirts execute quick allegro combinations with changing patterns, rather in the style of Balanchine. A young man and woman do an *adage* together, followed by another couple. In a new scene, a woman *bourrées* in and then several dancers suspend themselves on hanging ropes. Four girls do scribble-and-doodle kinds of movements that are challenging and complex. Tudor emphasizes birdlike arm movements. The score is Nikolais-like.[55] Tudor keeps adding on and peeling off the different dancers. Once again six women playfully hang on the ropes. The stage darkens and the dancers come out to bow to one another. Two men and ten women bow in canon to each other. During the next scene, the women look up at the rope and the barre in dismay. The last dancer finally comes out to bow while the stage-shy gorilla makes a last entrance, playfully jumping on the rope.

Passamezzi (March 8, 1962)

This five-minute mini-ballet was created for the *Gradus ad Parnassum* (1962) Ballet Studies performance that included *Dance Studies (Less Orthodox)*, *Trio con Brio,* and *Little Improvisations.* . . . In the case of *Passamezzi,* one might ask the question, How short is a ballet, or should a ballet be? Baroque music by Antonio Gardano with a Spanish flavor accompanies this brief but elegant piece with a strong sense of period style. Upstage there is a Moorish shaped arch, downstage a bench two young men and a young woman are sitting on, their backs to the audience. The dance seems to include ideas from his other short pieces. The genteel young men carry a cloth, and in turn, each of them does a patterned, baroque-inspired dance. They wear white shirts and grey tights. The young woman, in a white, short tunic skirt, dances with them. When she sits back down on the bench, each man comes out as if to competitively woo her. Tudor plays with the baroque dance forms. The cloth first hides a bouquet of flowers, then a medieval orb, a symbol of office that represents the kingdom. Finally the cloth conceals a skull, which the young woman chooses, adding a sardonic touch to what appeared to be an innocent game. The young woman finishes under the arch with her young man (who presented the skull) in a lunge looking downstage. Tudor makes no attempt here to test the virtuosic mettle of these dancers. Rather, he concentrates on the carriage of the body, gesture, and rhythmic sensitivity. In this interesting and musically complex piece, Tudor used a minimal and compressed dance vocabulary.

The years from 1950 to 1962 are often seen as Tudor's middle-age years of failure, of "movement silence." On closer inspection, the truth is more complex and more troubling to the dance historian. At the age of forty-two, Tudor created his last work for Ballet Theatre, *Nimbus*. He spent the next years as an itinerant dancing master, his only dependable income being from his dance classes at the Metropolitan Opera Ballet School and Juilliard, where Tudor established an important teaching methodology of a more personal and intellectual understanding of ballet movement. But despite heavy responsibilities to his students, Tudor never abandoned choreography. In fact, this period in his career began with an exuberant burst of creation, as though his final divorce from Ballet Theatre had inspired him.

Tudor's very real choreographic silence between 1955 and 1958 may have been caused by the bitter realization that whatever he made might be likely to sink in ignominious failure, since both the critics and the public seemed permanently disgruntled with him. Nothing Tudor created recently could match their expectations: "This was no *Pillar* or *Undertow,*" they continually complained. His smaller works designed around the strengths and limitations of his students and presented in minor venues were simply ignored or dismissed by the important dance critics.

Ultimately, there is no need to look for deep psychological motives for Tudor's "silence." A major ballet cannot be made easily without a major dance company and the commitment of major funding, two requisites Tudor could not manage to promote for himself in those frustrating years. Perhaps what is amazing is not Tudor's twelve-year banishment from critical favor, but his refusal to fade away.

In the years from 1959 through 1962, he gradually and painfully—without the support of a ballet company devoted to performing and preserving his many great ballets—re-established his career and broke back into the highest and most serious reaches of the international ballet scene. It is not entirely surprising that he had to return to Europe to do it.

CHAPTER EIGHT

Tudor Extends Himself

*I*n the early 1960s, Tudor's personal life became more involved with the First Zen Institute. In addition to the comfort of living with people he prayed with every morning, he found the quiet, austere environs a perfect place to live. Mary Farkas said that it took Tudor a while to familiarize himself with Zen concepts and practices. He became a senior member, an elder, in January 1960. From the time he moved into the Institute in 1962, however, he integrated the religion into his life so that he and the people within the Institute lived Zen together. He became more deeply involved with the Institute and was named president in 1964. Just as Tudor could bring sensitive emotional feelings out of his dancers, so he was able to encourage or awaken the spontaneity and intuition of his friends at the Institute. Farkas recalled that Tudor was almost too concerned with upholding certain Zen practices and traditions. For instance, he wholeheartedly embraced the ideal of having very few belongings, and he discarded almost everything, including many of the letters, records, and memorabilia of his career. It was a choice that left his future biographers with less information. But as a result of this Zen way of life, gradually the crueler and more bitter side of Tudor's personality quieted down in the studio. Farkas revealed that, when asked by his accompanist, Elizabeth Sawyer, why he had changed, Tudor responded, "I handle it all better!"[1]

As he found personal grounding from his connection with the Zen Institute, Tudor also gradually found a new creative energy. The movement "silence" that had plagued him in the nineteen-fifties gave way to a more productive period in the 'sixties. With Tudor's appetite for travel and exotic interest in discovering new places to work, he voyaged in June 1962 to Tel Aviv and Jerusalem, where he spent a month teaching. He mentioned that, while in Israel, he had a touching experience with the dancers. They were hungry for outside influences and especially grateful to work with a distinguished choreographer. But they also needed extensive attention to line and style, not having had intense backgrounds in the classical technique.

That summer, in 1962, Tudor traveled a good deal in Europe, and on August 1, he re-created *Jardin aux Lilas* at the Folkwang-Ballet of Essen. His former student Pina Bausch, who danced the role of Caroline in this production, arranged for Tudor to work with the company.

Tudor returned for another short period to the Royal Opera House, the Kungliga Teatern in Stockholm, Sweden, in September 1962 to reassume the directorship of the Royal Swedish Ballet and to set several ballets for them.[2] After rehearsing the company during the fall, he put on his *Romeo and Juliet* on December 30, 1962, as well as *Pillar of Fire*. *Romeo and Juliet* was unsuccessful, while *Pillar of Fire* stunned Stockholm, especially Mariane Orlando's understated but steamy performance as Hagar.

The company's repertoire now contained these ballets as well as *Dark Elegies* and *Gala Performance*. Soon after the performances, the Swedish dance writer Anna Grete Stahle, wrote:

> The Tudor programs were appreciated, but they did not attract large audiences. The reason is perhaps that his ballets have the flavor of a past period. This was felt at the revival of *Romeo and Juliet*. . . . *Pillar of Fire* has an honored place in the repertory of the Royal Theatre . . . it needs time to establish itself. The performance is not satisfying today, not sufficiently rehearsed or danced; but it has great possibilities.[3]

Tudor was not discouraged; on the contrary, this was a loyal and dedicated company, which he felt deep attachment to. With serious drilling and a new attitude toward the dramatic interpretation of movement, he was able to transform their dancing and their performance knowledge.

When Tudor returned to New York, he continued his classes at Juilliard and the Met, and in addition made plans to give a series of lectures at the New School for Social Research. They included sessions on the craft of choreography, as well as dance history, with guest speakers. The sequence of the lectures developed chronologically: "Noverre, the Great Innovator," "Bournonville, the Source," "The Romantic Period," "The Classic Ballet," "Fokine (the Rebel)," "Diaghilev (the Early Period)," "The Diaghilev Influence," "The English Choreographers," "The Break with the Classic Tradition," "The Contemporary Scene." It was a fascinating course on the evolution of ballet.

In these years, Tudor continued working intermittently with the Metropolitan Opera Ballet dancers. *Hail and Farewell* (1959) did not receive followup performances; but at the beginning of the new year in 1963, Tudor started rehearsals for a wonderfully silly and charming ballet, *Fandango*. It was in the manner of Spanish dancing with a sideline that parodied opera divas. The ballet was prepared for a program performed by the Metropolitan Opera Ballet at Town Hall in New York City. On the same program at Town Hall were the Rose Adagio from *Sleeping Beauty;* a new ballet, *One in Five,* by Ray Powell; and excerpts from *Les Sylphides,* directed by Alicia Markova. Markova was soon to become the artistic director of the Met Ballet Company.

Fandango (March 26, 1963)

Fandango has the studied grief and grotesque charm of a Goya painting. It evolves very realistically as a dance competition between five women. Tudor chose ballerinas whom he felt a singular rapport with as they understood Tudor's intent for the ballet.[4] Descriptions of each woman's character may be found in the notation score. Their names are: "Desideria, the sexiest of the group, Nana, the hypocritical woman attempting to be a lady, Conchita, the flighty and giddy one, Esmeralda, the toughest and most aggressive one, and Serafina, the only true lady of the group." Tudor also assigns a perfume scent to each woman as well as a different headdress.

The ballet, to a baroque harpsichord *cum* piano score "Fandango" by Antonio Soler, is costumed in Spanish style, with large, ruby-colored, lace-trimmed skirts to the calf, over layers of muslin. Their heads respectively carry a black mantilla, a rose by each ear and a black snood, a white mantilla, a black Spanish hat, and a rose on the right side of the hair. Included in the notation score is a description of the setting:

> A public square in the South of Spain, with general overall lighting. The five girls meet in the square and immediately start vying for attention. Each one thinks that she is better than the others and tries to prove it. This rivalry or competition is serious without becoming vicious, and even playful at times. Everything is done for and at the other girls. The dancers are not conscious of the audience.[5]

Thus they all compete in a kind of female dance-off. They begin to sing the music, arch their backs and pose their circular arms as they pound their feet to the intricate rhythms. There is a bench onstage, a common prop in many of Tudor's ballets, but especially appropriate to an event that has flamenco accents. Tudor invites each woman to enter with a little solo, quite individually developed, and then interweaves their dancing. They all clap in a circle as the solos ensue. The competitive quality adds a tension and edge to the driving rhythms. One woman even pushes another out of her dance

Fandango, *1968, with Juilliard dance students.* (Photo by Elizabeth Sawyer)

space. The hands are high, sometimes fingers clicking, then low, with quick-witted footwork, little *temps de cuisses,* hitting the back heel, jumps into second with *coupés* and swishing moves and kicking skirts. Even though wearing pointe shoes, the dancers are asked to replicate the footwork of traditional Spanish dancing. They do not always stay with the steady, driving beat of the music; they move on top of it, some with it, others against it, occasionally half-time or double-time. They do slashing footwork and stamping of the feet. One dancer collapses on the bench; makes a *thop* sound; she gets up and claps. When she does little footbeats, her hands and arms are held in a stylized manner, but they do not move very much; as one lady sings, they all parody her singing. Sounds seem to come from everywhere—they slap thighs, click heels, clap hands, snap fingers, and sweep their skirts with swishing sounds. Tudor emphasizes the quality of lace and filigree in their sinuous use of floreales.

Towards the end, they each do earthy, flatfooted jumps. Then in unison Basque dance steps facing the audience, they present an authentic image of village women who jump from elegance to true peasant origins; Tudor has his local color down. Soler's music perfectly drives these five beautiful hags; you can almost hear the witches call. Each woman both triumphs and is defeated in this captivating, but brief dance duel (twelve minutes).

Later reviews for the ballet have been positive, though initially it was given only two performances by the Metropolitan Opera Ballet Company.

Fandango has been revived by a number of dance companies over the years, including the Ballet Guild of Cleveland (1971), the Cincinnati Ballet Company (1971), Joffrey II (1971–1973), the National Ballet of Canada (1971–1974), the Louisville Ballet (1980–1983), the Dance Ring (Diana Byer, 1980–1983), the Hong Kong Ballet (1985), and the American Ballet Theatre (1988).

In the summer of 1963, Tudor traveled to Stockholm, where he restaged *Giselle*. While in Europe in July, Tudor also visited London, where he was invited to oversee a full evening of his works with the Rambert ballet. This was a big audience-pleaser. Mary Clarke in the *Dancing Times* reflected that "Ballet Rambert's 'Tudor Programmes' are excellent box office during the London seasons; they bring in, all on the same night, people who have nostalgic memories of *Jardin aux Lilas, Judgment of Paris, Soirée Musicale, Dark Elegies,* even *Gala Performance,* and young people interested in ballet who want to see the works which helped to establish Tudor's great reputation."[6] Tudor evenings were a regular feature of Ballet Rambert. The first full Tudor concert took place on September 11, 1958, curiously announced as an "American Evening"—presumably as Tudor had become so identified with the United States. The program consisted of *Lilac Garden, Judgment of Paris, Dark Elegies,* and *Gala Performance.* In 1962, the Tudor evening expanded when *Soirée Musicale* was added to his repertoire. This particular program was repeated in 1963 and 1964. The company re-formed in 1966 and became to a large extent a modern dance company. The Rambert Dance Company continues to dance Tudor works, however, some with beautifully inspired performances. Jane Pritchard declared, "Our *Elegies* is legend now!"[7]

Whatever the reasons for his renewed confidence and productivity, the fact is that Tudor went back to Sweden after his triumph in July in London with his creative juices flowing, and at the age of fifty-five he produced for the Royal Swedish Ballet company his first major new work in thirteen years.

Echoes of Trumpets, Ekon Av Trumpeter (September 28, 1963). Later retitled Echoing of Trumpets

Tudor worked on his new ballet in Sweden for a month. He vacationed in Rome, then returned to Sweden to work five more weeks on the piece. In a letter to Lucia Chase dated September 8, 1963, he writes from the Kings' Theatre (Kungliga Teatern) in Sweden that he "threw out the Tchaikovsky and the Martinu is giving me more difficulty than I could have possibly imagined and I am trying to accustom myself to the possibility of a 'still-born little bomb.' Everyone raved about the first ballet performance of the season last Thursday and so, Grace à Dieu for his small mercies, Love, Antony."

On the same evening as the premiere of *Echoing of Trumpets,*[8] Tudor restaged *Little Improvisations,* calling it *It's Raining.* Later, Tudor reminded American critics that, in Sweden, he had five rehearsals with the ballet orchestra for *Echoing,* which gave the dancers a chance to hear the music as it really sounds. Unfortunately in America, there is usually only one orchestra rehearsal (it is too expensive to have several); thus the dancers are forced to learn the music as they perform.

Tudor had taken a long time in his career to return to a serious theme that touched upon the war and violence; nevertheless he was the first choreographer in Europe to revisit World War II in balletic form. *Echoing of Trumpets* presents universal images of village women alone against a brutal occupying army. *Echoing* was viewed as a new kind of *Dark Elegies* dealing with unpleasant truths. "The characterizations here, like those in *Dark Elegies,* are generalized. The women are symbols of courageous devotion. The men are cruel in action."[9] When Jack Anderson interviewed Tudor, however, Tudor admitted that most audiences associate the ballet with World War II, and there are resemblances between the plot and the historical event of the destruction of Lidice, Czechoslovakia, by the Nazis.[10]

In the American Ballet Theatre program, November 30, 1967, the notes specify that "the theme of *Echoing of Trumpets* is closely related to another composition by Bohuslav Martinu, Memorial to Lidice, of which Mr. Tudor, by some strange coincidence, heard only after he had begun working on the ballet." In the interview with Anderson, Tudor recalled that "the Martinu music deeply touched him hearing at times moments of gun shots and plagues of locusts."[11]

Where possible, Tudor liked to use theatrical settings for his ballets. In *Echoing* he chose a harsh, jagged, multileveled design that emphasized a feeling of both entrapment and escape. There is only one place to enter and exit the stage.

Echoing of Trumpets, *1963. Antony Tudor rehearses Gerd Andersson and Mario Mengarelli with the Royal Swedish Ballet.* (Courtesy of Gerd Andersson)

The ballet begins on the higher level, where a woman oversees the surroundings, as if watching to protect the town. Other women, dressed in simple dresses, wearing shawls, carry on this sense of watchfulness. Fragile and tentative in their movements, the women have been abandoned, and there seems to be no protection anywhere. Tudor often raises the women on their pointes, moving them slowly in mannequin-like movements. They come together, arms stiffly held at their sides, huddling. The mood is one of desperation. Trumpets echo throughout the symphony. As Tudor recalled, he chose the Martinu music because as a child he heard bombs falling, and remembered the trumpets blaring from an army camp with a firing range

Echoing of Trumpets, 1963. *Royal Swedish Ballet.* (Photo by Beata Bergstrom)

near his home. The chorus of women move in closely guarded patterns of movements. When the soldiers pour over the wall, it becomes clear what the women were anticipating. Dressed in uniforms and heavy boots, the soldiers move in pounding and weighted steps.

Tudor has given them a different movement vocabulary and style. They do not do "ballet." They move as Greek or Eastern European folk dancers would, holding each other's shoulders in a strong and determined way. But there is nothing attractive or friendly about them, and that Tudor has captured from the start. They become a relentless force, frankly sexual in their attacks. When a woman brings bread to share (and this is a moment that does not really work for me), some women savor a piece while another tries to devour it. The morsel is brutally knocked from her hand; she drops down to grab for it, fumbling while the soldier smashes her hand into the ground. Here Tudor incorporated a story he was told about a Greek peasant reaching for a crust of bread and a Nazi crushing his hand into the earth. This encounter presages others of more desperate import.

The central moment of the dance is the *pas de deux,* a grotesque dance of death in which one of the women dances with her dead lover. During his reckless attempt to sneak back into the village to see his wife-lover, the soldiers string him up by his feet on a gallows at the peak of the back wall and execute him. Tudor places this scene at the highest point on the set, and the viewer focuses on the vulnerable verticality of a hanging body, upside down, "jerking like a fish on a line."[12] After the women cut his body down, the tormented wife dances with his dead body. She tries to pull it up onto her, to breathe life into it, to speak to it. She takes his clasped hands into hers and drags his limp body around in a circle. She then flings him away with such violence that one cannot but sense an anger that far exceeds any sorrow.

Once again the soldiers enter; the women plot revenge, and suddenly the shawls they hold become menacing, the means to strangle and kill one of the soldiers. Their retaliation echoes Jooss's *Green Table* when the Partisan strangles a soldier with her scarf, as well as Federico García Lorca's *Yerma,* when, during the last scene, Yerma uses her shawl to garrote her husband. This action brings down the final wrath of the invaders. In a dance of multiple *pas de deux* the women are tossed from one man to the other and are raised and stretched out as if on a rack. The men dance around the group of women in what is almost a male witches' dance. In the end there are only two women left, wearily keeping watch. They know their fate. The trumpets call once more, and the stage is dark.

After the premiere of *Echoing of Trumpets,* Anna Grete Stahle described her reaction to the ballet in the journal *Dance Perspectives:* "I am very glad to tell you that Antony Tudor after much hesitation has produced a strong ballet. It is not a spotless masterpiece, but the total effect is strong and well, violent. This may very well be the beginning of a second creative period."[13]

Clive Barnes spoke of the International Festival of Dance in Paris, where the Royal Swedish Ballet presented

> one of the best pieces of ballet news in many a long year. Antony Tudor, for nearly 20 years, nothing seemed to go right for him. *Echoes of Trumpets* looks like the long-awaited Tudor masterpiece. He has given us a profoundly anti-romantic ballet about war—a ballet that is real, terrible and yet still beautiful in the scarlet way of tragedy. It has all that poignant immediacy associated with Tudor's early work. Bitter, perhaps pessimistic, it is one of those ballets that have been hewn out of a human soul.[14]

Tudor discussed the meaning of his ballet with Jack Anderson: "Perhaps it's more about how people always seem to want to dominate other people. Everyone knows that's a stupid thing to do. Yet they keep on doing it. They never stop torturing each other with a kind of mild viciousness." Tudor added that "I've known some specialists in it . . . even in ballet studios." A wicked glint appeared in Tudor's eyes: "Take the soldiers in my ballet. They don't really rape the women in the village. They just torment them until they make the women feel degraded and, in so doing, they degrade themselves."[15]

With this explanation, Tudor broadened the message of his ballet. Yet most people who wrote about the ballet when it opened in Stockholm associated the scenario with the war and with a specific event in Eastern Europe, especially as the rape scene seems to lead to the women's violent end.

Tudor arranged the first American performance of *Echoing* to be danced by the Metropolitan Opera Ballet Company on March 27, 1966. The ballet was prescient in 1963 for America; but by 1966, it became a metaphor for America's relationship to the Vietnamese, with our soldiers behaving at times in a horrible way toward the native population. *Echoing of Trumpets* bears some resemblance to José Limón's *Missa Brevis* (1958), as both dances are set in Europe and are protests against the guiltless's being destroyed.

It was a pity that *Echoing of Trumpets* had only one showing at the Met, but the results were successful. After that performance, a number of ballet companies wanted to re-create it.

Echoing of Trumpets has been performed by Ballet Theatre (1967), the London Festival Ballet (1973), the English National Ballet (1980), the Deutsche Oper in Berlin (1984), and the Louisville Ballet (1990).

Immediately after the premiere of *Echoing of Trumpets,* Tudor traveled from Sweden to Berlin, where he staged *Gala Performance* and *Jardin aux Lilas* at the Deutsche Oper for the Berlin Festival Ballet.[16]

Although very few of his ballets had been requested by Balanchine's New York City Ballet since 1950, Tudor re-established a relationship with them in 1964. *Dim Lustre* was re-created at the New York State Theatre by the New York City Ballet on May 6. Tudor did not feel that the City Ballet dancers exhibited either the lyricism or the subtle expressivity that *Dim Lustre* demanded.[17] Indeed, John Martin also observed:

> The performance is not yet everything it should be. The company is primarily attuned to the style of George Balanchine, and there could scarcely be a style less like it than Mr. Tudor's. As a result, the texture of the piece has not been realized and it is on this that the dramatic continuity depends almost exclusively.[18]

Allen Hughes seconded Martin's reservations:

> The chances are that this 1964 *Dim Lustre* would have come out differently had Mr. Tudor done it with a company of his own, one that he had trained over a period of years.[19]

Tudor visited Japan for the second time in June 1965. There he restaged *Dark Elegies, Jardin aux Lilas, Pillar of Fire,* and *Undertow*. Ruriko Tachikawa was the producer of five evenings with the Tokyo Philharmonic Orchestra sponsored by the British Council in Japan. Tudor wrote a brief message in the printed program, thanking the Japanese for their invitation. "I accept this invitation in the hope that my work with the ballet of Japan . . . will greatly further and accelerate the development of a truly national ballet which in time may become a national pride and a valuable export."

Working with the Metropolitan Opera Ballet Company may have frustrated Tudor, but it also pleased him. He knew that their work did not conform to the way most ballet companies function—i.e., without the difficult hours, the commitment to classical ballet, and the strenuous performing schedule of a regular ballet company. Yet he found Met dancers' versatility with other dance forms, their acting talents, and their sensitivity to music useful skills in creating a new ballet. In the spring of 1966, Tudor made a new piece for the dancers at the Met.

Concerning Oracles (March 27, 1966)

Concerning Oracles began with an experimental idea at Jacob's Pillow during the summer of 1951. *Les Mains Gauches* told the story of a couple and the character Fate as they play out the more sinister angles of being a left-handed person, starting on the left foot, heading leftwards, and so forth. In palmistry, the right hand signified what you are born with, and the left hand

Concerning Oracles, *1966, with the Metropolitan Opera Ballet.*

offers insight into the future. On this March 26 evening, Tudor's *Echoing of Trumpets* and the historically important August Bournonville's *La Ventana* had their premieres at the Met. *Concerning Oracles* also had its premiere on the same evening and was, perhaps, the least successful of all the ballets.

In *Dance Magazine,* Tudor explained the origins of the ballet. "It grew out of the Ibert record with the tantalizing pieces of music on it and the sheer necessity of having to do a ballet to fill out the Met program."

Sallie Wilson suggested that he revive *Les Mains Gauches,* but Tudor felt that it was too slight a piece for the huge Metropolitan Opera House. Tudor decided to include it as a part of a larger work about crystal-gazing and card-reading:

> People tell me the ballet is obscure. I don't know why that should be.
> . . . Everybody knows what fortune telling is like and what its problems are.
> It's inconclusive. When you have your fortune told, do you believe it? Does
> what you are told will happen to you really happen? No, I've never been to a
> fortune teller myself. I never thought it was necessary.[20]

In the journal *Choreography and Dance,* Sally Brayley Bliss recalled that they were rehearsing *Echoing of Trumpets* and *Concerning Oracles* at the same time, and had to "jump from one work to the other."[21] Tudor had a number of dancers learning more than one role in the ballets, and up to one week before the show, he had not officially cast them. While working on *Concerning Oracles,* Sally Bliss remarked:

> The pas de deux was so difficult . . . every lift, every combination was to look as if it came from nowhere. You were never to show any preparation to go onto pointe or into a lift. In order to accomplish that quality, we used muscles we didn't know existed. Tudor would demonstrate what he wanted in his street shoes, shirt, tie and pressed pants. We would argue, "You're not in pointe shoes, Tudor, it's not possible." He would say, "Do it," and we would.[22]

Furthermore, Bliss described how Tudor worked with Lance Westergard and herself in another *pas de deux* scene:

> I was a dowdy old maid aunt. Lance was my nephew, with dreams and imagination. As he came to my part of the table, I stripped off my dowdy dress and appeared in a beautiful sexy pink peignoir. We started to waltz. As the music started to build, the obvious thing was for Lance to lift me. The choreography would build to the expected lift, but at the very last minute, I would lift him. The pas de deux kept building and the lifts were harder and harder. It was absolutely hilarious.[23]

Concerning Oracles is composed of three episodes, both sinister and comic, concerned with the gifts of prophecy. In the first episode, an Elizabethan lady thought to be Mary, Queen of Scots, confronts the powerful symbols of kingdom (an orb), marriage (a chaplet), and death (a skull).[24] The second episode, "Les Mains Gauches," deals with the rose and the noose as possible choices during a *pas de deux.* Though the original character of Fate was changed, the atmosphere of hidden horror still pervades the dance. The third episode, "L'Arcane," with its setting of a family outing at a picnic table in nineteenth-century provincial France, was a prophetic comedy in which an imperfect fool becomes involved with Tarot cards, love, war, and marriage.

Clive Barnes declared that "the schematic nature of the ballet was given continuity by the link-figure of the Fortune Teller, although the blandness of the choreography and the fractured tri-partite score proved too much for the work's unity."[25] Walter Terry also panned the ballet, except that he appreciated both the scene with the Tarot cards and the frisky, charming duet between Lance Westergard and Sally Brayley.

A. V. Coton delivered the most penetrating and descriptive review of this evening at the Met. Because of the presentation of *Echoing,* he reassured the reader that it was largely a triumph for Tudor:

However *Concerning Oracles* showed the faults of a choreographer too long absent from ballet-making. The situations were too slow in development, the secondary characters, too imprecise. Yet it reveals Tudor's unique capacity to create human-sized characters through classical dance movement undistorted by naturalism. Each of the three episodes from different historical eras, Elizabethan, Romantic Period and Maupassant Pastoral centers on people's reactions to the impact of prophecies. The triple action is linked by a fortune teller dispensing spells, gruesome symbols and Tarot cards. The first two incidents carry overtones of wonder and horror, while the third explores the furiously comic yet unhappy adventures of a gauche youth seeking the admiration of a lovely woman. . . . The three episodes do not create a bold triptych commemorating the behaviour of victims of supernatural possession. Tudor's alchemy fails in its total effect because this group of technically able dancers are, on the whole, insufficiently sophisticated to give real depth to his sardonically imagined characters.[26]

Possibly because the Metropolitan Opera Ballet School did not figure in the plans to move the Metropolitan Opera Company to Lincoln Center in 1966, Tudor decided to quit directing the School. During the same year, he received the honor of an unrequested National Endowment for the Arts grant for $10,000.

When Tudor was invited by Frederick Ashton, the new director of the Royal Ballet in London, to do a ballet for them in June 1966, he must have been both delighted and surprised. The fact that Tudor had not been encouraged to stage *Pillar of Fire* for Margot Fonteyn in the 1950s or 1960s reflected the English reluctance to recognize someone who did not fight alongside them in World War II.[27] Though Tudor never publicly accused Ninette de Valois of slighting him, there is no question that she deliberately excluded Tudor's choreography from the Royal Ballet company. Finally Ashton asked Tudor to create *Shadowplay* for the Royal Ballet. Was it a coincidence that Ninette de Valois had stepped down as director in 1963?

That fall, he went to London to begin rehearsals of *Shadowplay,* which was completed and performed in January 1967.

Shadowplay (January 25, 1967)

Shadowplay does not resemble any of Tudor's other ballets and reinforces his own dictum that he never liked to repeat himself. Also, he experimented with a movement vocabulary that seemed inimical to the cool, carefully positioned, and fluid English style, thus testing his countrymen with, not only a challenging scenario and a trapeze, but also a vastly different technique.

Shadowplay, *staged at Ballet Theatre, 1975, with Mikhail Baryshnikov and Jonas Kage.* (Photo courtesy of L. D. Vartoogian)

Tudor had been thinking for several years about a ballet with dancers swinging from ropes. He used a circus set for *Dance Studies* (*Less Orthodox*) (1962). Remy Charlip wrote a review of *Shadowplay* and remembered that, while he was a student at Juilliard, Tudor put up the trapeze in the dance studio: "He created a piece in which the dancers fooled around with one another and some were suspended on bars. Then an ape lumbered in, swung up onto a bar above them, watched for a while and scratched his head. Curtain. 'We are monkeys all,' Tudor seemed to say."[28]

Antony Tudor had an abiding interest in Oriental religions and literature for many years. In his *Atalanta of the East,* Tudor was lured by an Indian interpretation of a Greek myth. In 1948 he created the calamitous *Shadow of the Wind* for Ballet Theatre based on the poems of Li Po and other Chinese writers to the unlikely musical score *Das Lied von der Erde* by Gustav Mahler. A year later, in 1949, Tudor choreographed a short duet for Hugh Laing and Diana Adams at Jacob's Pillow, *The Dear Departed,* about a lady genie who pops out of the empty bottle of a wistful and thirsty Chinese peasant. If

209

Shadow of the Wind did not bring Tudor positive reviews or artistic satisfaction, *Shadowplay*, his experimental adventure into Indian Buddhism, fulfilled his hopes for a moderate and interesting success. It also provided exciting research into the Buddhist religion, as his profuse notes testify. Years later in an interview with Zita Allen, Tudor confessed that *"Shadowplay* is a Buddhist ballet in disguise, of course."[29] It might be suggested that Tudor's decision to use an Oriental theme came with the understanding that his countrymen in England had a longstanding fascination and affinity with their Indian commonwealth.

Tudor discovered the music of a little-known Alsatian composer, Charles Koechlin (1867–1950), who was trained by Gabriel Fauré and was a disciple of Debussy's. Koechlin tended towards impressionistic themes. He took delight in contemplating nature and then translating his impressions into sound. Although there is no synopsis of the ballet, the program for *Shadowplay* explained that Tudor chose the main structure of his ballet from Koechlin's "Les Bandar-Log" (1939). All the original score is used, but one section is repeated after the first of the two interpolations. The added material is not extensive but is derived from a much larger symphonic poem, "La Course du Printemps" (1925–1927). Both compositions belong to Koechlin's vast fresco *Le Livre de la Jungle,* based on Rudyard Kipling's *The Jungle Book* (1894).

For both Koechlin and Tudor, the jungle represented a revelation of the mystery of things and beings. The "Bandar-log" are the monkeys. Koechlin treats them as symbols of anarchy and vulgarity—in opposition to the order and the mystery "of things and beings" represented by the jungle and its noble creatures and by Mowgli (the hero or anti-hero) himself. Tudor prepared himself for rehearsals by jotting down extensive notes culled from Koechlin's score and various Buddhist and Hindu writings.[30]

In addition to tempo markings, Tudor recorded his studies of the Hindu gods, many of them animals; they provide a rich background for the animal characters who surrounded Antony Dowell in his role as the Boy with Matted Hair. Specific images such as "tightrope dancer who slips" and "naked streaked with ashes" combine with attributes and qualities such as "pendulous between twoness and oneness," "emptyness finds emptyness." Tudor's jottings describe Apsaras, courtesan dancers of heaven, consorts of the Gandawas, spirits of mid-air who are uncommonly beautiful: by their languid postures and sweet words, they rob those who see them of their wisdom and intellect. These Apsaras serve as models for the sensual creatures who come to haunt the Boy with Matted Hair. The god Siva who inspired the character, the Boy with Matted Hair, is represented as the perfect ascetic. "He is shown naked, clad in space, loaded with matted hair.

He is the Dance of Death—Siva can only unite with himself, for perfect beauty can see only itself reflected in itself." Tudor's notebook continues:

> In Chod, a human being mystically attempted to realize salvation in himself by means of ritual symbols and symbolic actions. While each step of the dance had been carefully learnt, he lets his head be cut off and his body devoured while remaining fully conscious, so that as a result, he was free of his ego and found a "point of rest." He may go mad, but Enlightenment is a precious gem and must be bought at a high price. He dances the dance that destroys erroneous conceptions—creates, destroys, recreates and again destroys in a wild dance uniting sexuality and aggressivity at their deepest level where they are complementary to each other as two aspects of the same power.[31]

The woman as evil temptress in Buddhist mythology plays a significant role in Tudor's marginal notes for this ballet. But the woman as a violently treated being also appears. "On a lower level, this female combination of lasciviousness and lust for destruction could be raped by the Yogi, i.e., conquered and made subservient to his will (heroic incest)." At a higher level, she was the image of the dreadful sensory world of nature, the bodily sensuality and self-destroying lust that cause human beings to believe in the illusion of an "I." The issue of ego and identity fuels much of the action in *Shadowplay*.

All of these curious notes call attention to Tudor's intense research; they provide, to some extent, a dictionary of terms that explain his vision of *Shadowplay*. What seems apparent and important to Tudor's ballet is his deep connection to the layering of human experience. Gradually, one will become pure, gradually one will learn the ways of the world, gradually one will denude oneself of sin, pride, and greed. Levels of innocence and goodness are experienced. The jungle setting tends to obfuscate and make more difficult this peeling-away. The confusion and shadows cast by the trees and thickness camouflage the truth. One must search. There is no question but that women are the enemy. Here is a mysterious mythology of humans and beasts where huge birds, dragons, monkeys, and insects interact with or become humans. The masculine does not seek the feminine as in the animal world. Rather it encompasses it or becomes it. Tudor's concern here is enlightenment or *satori*. "He may go mad but enlightenment is a precious gem and must be bought at a high price." "The dreadful sensory world of nature" causes man to believe in the illusion of his ego and it must be destroyed. One must transcend the limits of one's time.

The theme of the ballet is growing up, or "a man's progression to a state of nirvana beyond the distractions and irritations of the world. The young boy is beset by the menaces of the jungle (the world) and its creatures of the trees and of the air. He is confronted by a male figure (Lord of the Jungle or his earthly self). He is successively charmed and threatened by the Celestial (a chaste goddess or seductress). In the end, he achieves his peace (manhood) through an act of will (or the sexual act)."[32] In an article, "The View from the House Opposite," Jack Anderson found Tudor "slightly scandalous as both the scenes with the woman and the guru possess sexual implications, the guru seeming both a homosexual lover and a master."[33]

The ballet's major figure, the Boy with Matted Hair, has his origins in the *Jungle Book*'s Mowgli. But Tudor stretched this paradigm (his costume attests to the fact) to be the "embodiment of all young manhood about to take his first steps."[34] The Penumbra, the space of partial illuminations between the perfect shadow and the full light, is divided among the Arboreals, the Aerial Terrestria, and the Celestials, "distractions that impede our deeper contemplation about life."[35]

The ballet begins where the Boy with Matted Hair remains alone in the jungle, in the womb of his existence where he examines the world of nature and his consciousness. Contemplative balances and slow turns characterize this meditation. Suddenly tranquility is interrupted as our ancestor monkeys cavort around the forest and cajole the Boy. They seem to take pleasure in their discovery of him. Birdlike "Aerials" fly in with exotic plumage, dressed as Cambodian dancers, and an imposing and most interesting Terrestrial, a powerful, dominating man, made to look like a Hindu deity, begins to joust with the Boy. One senses a very important moment. The Boy slips through the Terrestrial's legs, down his back; until eventually the older man sits under the tree. Eventually the Terrestrial leaves, but we are sure that he will return. This encounter presages the Boy's knowledge of one person sexually overwhelming another. The "Arboreals" return to the Boy's jungle menacing and threatening him until he hides in the trees. There another threat, perhaps more terrifying than any other, descends on the Boy. The "Celestials," led by Merle Park, are attractive, sensual, and dangerous. Park and the Boy move together in a passionate coupling. "Celestial is borne aloft at the front and dips and swoops over him in a passage that is taut with drama, full of sexual undertones. . . ."[36] Their coming together takes him beyond his understanding. "They enter into an elaborate adagio in which she inflames him by crouching on his back, then is supported in slow pirouettes, but three times is lifted away from him then returned for more."[37]

This duet contains "one of the most astonishingly beautiful moments in the piece when they spin separately—it seems like triple turns—stop at the same time facing each other in second position and hold on for dear life."[38]

The major scene of the ballet occurs when all the creatures of the jungle unite to form a huge, snakelike structure; the "Celestial Being is being carried, her parted legs are supported by two lines of Arboreals. The scorpion-like phalanx, with the Terrestrial as tail, approaches the Boy who dives into the tunnel they form."[39] The Boy has achieved some kind of knowledge, be it sexual or spiritual. Revitalized and re-empowered, the Boy succeeds in forcing them to leave the jungle, his place of conquest. Eventually the monkeys scramble back and the ballet ends with their acceptance of his quiet mastery.

One might be concerned that the movement vocabulary in *Shadowplay* would border on the absurd, with monkeys jumping around everywhere. Yet Tudor's movement choices are subtle and suggest animal qualities rather than obvious pantomime, and his introduction of poses and gestures from Hindu or Cambodian dancing is done skillfully.

The English critics delighted in the ballet, and enjoyed deciphering the various scenes of confrontation. Clarke thought it a minor but beautiful work. Percival saw it as "richly suggestive, provoking ideas about the human condition."[40] Richard Buckle, also enjoying it, found it "just on the razor's edge, almost nonsense; Tudor hit it just right. The animal kingdom is with us for keeps."[41] A. V. Coton concluded: "No lesson is offered, no wrapped-up conclusion is provided; we are watching, perhaps, a human struggling to understand certain shadowy beings present only in his imagination. This is truly a shadow play, with its part amusing, part solemn encounters and entanglements taking place in that vague region that belongs neither to darkness nor light."[42]

The Royal Ballet brought *Shadowplay* to the Metropolitan Opera House in New York on May 3, 1967. Clive Barnes reviewing the performance wrote: "The choreography is beautifully apt and untroubledly imaginative, going smoothly with the music, upon which it seems to provide a physical and emotional commentary."[43] Anthony Dowell's performance also received acclamation.

The ballet was restaged at American Ballet Theatre for Mikhail Baryshnikov at the New York State Theatre on July 23, 1975. Gelsey Kirkland played the Celestial and Jonas Kage the Terrestrial. It was thought to be a good vehicle for Baryshnikov's acting talents. But Arlene Croce in the *New Yorker* magazine, never a great fan of Tudor's, caustically called the ballet "this rusting hunk of junk jewelry that was acquired for Mischa."[44]

In 1968, Tudor invited Steven-Jan Hoff, a young Dutch dancer, to play the Transgressor in a revival of *Undertow*. After its premiere in 1945, Tudor could not reconstruct *Undertow* at Ballet Theatre, unless he found a Transgressor who was young and fit the part; someone who was short, tightly built, and an excellent dramatic dancer. However, with Steven-Jan Hoff the 'sixties seemed a good time to replay the ballet, with its radically sexual implications. Toni Lander danced the dual roles of Cybele and Medusa, while Sallie Wilson enacted Volupia. With changes in the mores, perhaps moments in the ballet had less power to move the public. At any rate, it enjoyed only a slight success.

Despite the excitement of politically activated artistic "happenings," especially in the modern dance world, the war in Vietnam, and minority and women's rights evaded Tudor's choreographic interest. Rather, he occasionally alluded to societal changes, changes not necessarily limited to American youth, created by these events and their reverberations. For example, the seeming licentiousness and chaotic coupling, often caused by drugs, that characterized the 'sixties generation rather shocked Tudor. He said so a number of times, and two of his ballets, *Knight Errant* (1968) and *Cereus* (1971), certainly made oblique if not direct reference to this situation.[45]

Knight Errant (November 25, 1968)

One might wonder why, after the significant achievement of *Shadowplay,* Tudor was asked to create a piece for the lesser touring company of the Royal Ballet; the tour opened in Manchester! Certainly Tudor's feelings might have been hurt.[46] Seemingly unperturbed, Tudor created a jaunty ballet about sexual morality, wrapping it in an eighteenth-century scenario from the novel *Les Liaisons Dangereuses* by the much-touted Choderlos de Laclos (1782). This eighteenth-century mannered morality play was meant to be a comment on the 1960s' sexual revolution.

Tudor admitted that his ballets fell into two strains, one serious, one comic; the comic strain often mocked the sexual practices of his leading characters, just as his earlier *Gallant Assembly* (1937) parodied the libidos of several gallant ladies and gentlemen.

Knight Errant was based mostly on one episode in the Laclos novel *Letter 79,* that of Prévan (not Valmont) and the "celebrated and salacious affair which separated 'the inseparables.'" The letter describes a typical eighteenth-century scientific, rather detached approach to lovemaking and its consequences among acquaintances. To this Tudor added an epilogue based on Prévan's discomfiture (*Letter 85*) by the Marquise de Merteuil. The Woman of Consequence is inspired in part by a mixture of the Marquise and the beautiful foreign mistress whom Prévan deliberately quarreled with in order to have twenty-four hours to devote to the undoing of the "inseparables."

Knight Errant is characterized by several paradoxical contrasts to its scenario; for example Richard Strauss's overripe romantic, energetic music, a set and costumes with 1920-ish or 1930-ish abstract marble design, and white as the unlikely shade for all three of the Chevalier d'Amour's lovers. These disparate elements are set against the scandalous situations and erotic poses with a classical formality of line and eighteenth-century social dances.

The ballet opens upon the Chevalier d'Amour, a virtuoso of deceit, and his Woman of Consequence, with whom he shares a common interest in lubricious activities. Their relationship seems to center on voyeurism; they watch each other's sensual doings surrounded by Ladies of Quality and Gentlemen of Standing. Three inseparable Ladies of Position enter, followed by their protectors, the Gentlemen of Means. An atmosphere of general licentiousness is created by all of the ensemble. When the Gentlemen of Means search for further adventure, the Chevalier d'Amour moves in to the conquest of the three ladies left in the lurch, what might be termed "a sexual *Gala Performance.*"[47] The rake's first triumph is accompanied by a lush Strauss waltz, his prey a sweet, demure, protesting charmer. He brings her to the bed, undresses her, and all proceeds as planned. The Woman of Consequence appears with her own lover and leaves. The second Lady of Position's passion matches his own, though exhaustion should be setting in. Their *pas de deux* steams away with brief interruptions by the crowd. A totally comic third *pas de deux* ensues, with the hungry and immodest third lady wrapping her legs around his neck. The funniest episode occurs when the rake receives his come-uppance in his seduction by the Woman of Consequence, and she throws his clothes out the window, leaving him quite vulnerable and alone. He is shown up to the Gentlemen of Standing as well as to the Ladies of Quality. And the Ladies of Position are embarrassed to have been found out. In the end, the rake makes one final minuet gesture ("So what, there is still tomorrow") that leaves the audience wondering if he learned anything. He has had a rollicking good time losing the Woman of Consequence, a Mozart-like finale that combines pathos and wit.

Several critics objected to the changes of scene that were initiated by six postillions, each bearing an L-shaped frame that provided different set designs. They were reminiscent of the onstage curtain that initiates scene changes in *Romeo and Juliet.*

A. V. Coton, who saw *Knight Errant* in Manchester, asserted that this ballet of enormous vivacity "fascinates by the originality of its several parts, the theme, music, decor, or dance style . . . But the ballet is overlong in its development, imprecise in dramatic structure and often weak in choreographic design."[48]

John Percival disagreed. He praised David Wall's remarkable understanding of the role because he imbued it with a "casually impudent perfection. First, he inspires all the other characters. . . . Secondly he modulates the unexpectedly moral epilogue into the same key as the preceding hectic comedy." Percival enjoyed the ballet's wicked and bawdy comic qualities. He also noted that the ballet grew "more impressive on better acquaintance."[49]

Another assessment that connected Tudor's brilliant choreography with dancers who inspired him was written by Peter Brinson: "Few ballets illustrate more clearly Tudor's habit and need to create ballets around one artist or at most a very few for whose talent he has a special sympathy. In this case the artist is David Wall."[50] Richard Buckle spoke of it as a "wonderful performance of an extraordinary work."[51]

Undoubtedly, Tudor was elated to be back on home soil, to work with dancers of a similar sophistication and mindset. There were no loud complaints about Tudor's dilatory behavior.

Unfortunately *Knight Errant* has been lost as a result of not having been played. Perhaps notation or a secreted film will save it from permanent burial. This sad demise is all the more regrettable as its notices became more positive and appreciative with time. In Judith Cruikshank's article in *Dance and Dancers,* she remarked: "It really is disgraceful how the Royal Ballet has treated its Tudor repertory and having lost *Knight Errant* forever, let us hope that *Shadowplay* is better cared for."[52]

The following year, Tudor went off to Australia to work with the Australian Ballet Company, which was directed by his old colleague Peggy van Praagh. He looked forward to developing a piece that made connections to their aboriginal population. Instead, he made a ballet that was unappreciated down under. In fact, it seemed to be a colossal flop. While in Australia he also re-created a more successful *Pillar of Fire* on July 11, 1969, at the Majestic Theatre in Sydney, Australia.

The Divine Horsemen (August 8, 1969)

The opening night program indicated:

> The "Divine Horsemen" are those spirits or loa who may temporarily displace the conscious self of some person and manifest themselves through a living body. In the Caribbean area this process, which is commonly known as "possession" is said to be similar to that of a horse and its rider; and the loa in the "mounting" of a person takes over conscious power and control. In "possession," consequently the person who has been mounted cannot and does not remember anything that happens during this period for it must be understood that the self must leave if the loa is to enter as one cannot be man and god simultaneously. Erzulie, Ghede and Damballah are only three mem-

The Divine Horsemen, *1969. Members of the Australian Ballet.* (Photo by Hugh Fisher)

bers of a large pantheon, each of whom is archetypal and whose presences are invoked at communal gatherings. *Divine Horsemen* was the title of a book by Maya Deren and the choreographer is consequently indebted to his friend, the author.

Both the book[53] and the film that Maya Deren worked on from 1947 through 1951 in Haiti provided Tudor with further descriptions of this ritual drama. The film especially offers live footage of a particular village celebration. One is struck by the fearful energy and sacred power of the Haitians, especially when divinity becomes manifest during possession, or when the loa mount the person. Tudor and Deren had previously collaborated on the film *The Very Eye of Night.*

In an interview with Millicent Hodson, Vévé Clark, and Catrina Neiman for *The Legend of Maya Deren* project, Tudor referred to the coincidence of Alvin Ailey's bringing his company to Australia just a short while before the premiere of Tudor's ballet. In addition, Tudor speaks about *The Divine Horsemen,* confirming that Maya Deren's book of the same title was the inspiration for the ballet:

> I used Ghede, Erzulie and those people. I thought I was going to Australia to do a marvelous Alvin Ailey ballet and it would be the first in Australia; all his marvelous movement; and so I did it. I got a score, *Caribbean Themes,* by Werner Egk, of which I hadn't heard a recording and it sounded quite different in the orchestra from what I imagined from looking at the score. But it had some good sections in it. . . . Unfortunately, Alvin had been an enormous success with his movement a few months before I got there, which I never knew. I wanted to call him a "son-of-a-so-and-so." I thought I was going to show this to the Australian Aborigine; I was all wrong. You can't win.

Hugh Laing created a colorful, dynamic set for the ballet and carefully assisted Tudor in the development of the production values and the scenario. The set had several tiers that gave height to the stage, and painted flats with flowers, benches, and chandeliers in sunny Caribbean colors. The women in toe shoes wore full, layered skirts with ruffles and head scarves, while the men wore short trousers and open shirts. This was very much a theater play, a "ritual drama."

Photographs of several scenes suggest that Tudor's movement style was inspired by African Congo dances. The toe shoes seem out of place. Constance Cummings in *The Courier Mail* extolled the ballet "for the illusion of spontaneity created by Tudor's inventiveness." She called the ballet "the masterpiece of the evening." *The Melbourne Age* described the mood: "Extraverted people surrender to the influence of strange gods in an atmosphere of Voo Doo magic. The three possessed figures become gods from the Black Pantheon and perform their orgiastic dances with frenzied intensity in a dark setting dominated by a phallic symbol and urged on by Werner Egk's strident music."[54]

In his book *Ballet in Australia: The Second Act, 1940–1980,* Edward Pask, criticized Tudor's choice of story for the Australian Ballet company:

> *The Divine Horsemen* involves three people who are possessed by spirits or gods—a type of Voo Doo. One feels that this theme would have been better executed by the Dance Theatre of Harlem or Alvin Ailey. Only rarely do whites have that inner restless energy and rhythm which is associated with colored dancers and only in Karl Welander did anything of the frenzied dance come close to the locale of the ballet's theme.[55]

These remarks represent challenging attitudes toward dance and ethnology that are currently being debated. However intriguing this scenario seemed to Tudor, it probably drew Tudor too far from his understanding of the classical technique. After all, it was Tudor who admitted early in the 1940s that he did not feel comfortable with jazz themes or steps, as hard as he tried.

To our knowledge, the ballet has never been filmed or restaged.

On his way back to America from Australia, Tudor visited his brother Bob, a forest ranger who lived in New Zealand. They had had very little to do with each other for the last thirty years of Tudor's life, and one wonders what the choreographer and the forest ranger had to say to each other. However slight the contact with his family, Tudor's loyalty to them was strong, especially after his mother died. In his will, he left a tidy sum to his grand-nieces and -nephews.

During the fall 1969 season, Sallie Wilson recreated *Gala Performance* at Ballet Theatre, which received a none-too-favorable review from Arlene Croce calling *Gala* "over genteel and over general."[56]

The following year, Tudor received another $10,000 National Endowment for the Arts grant to assist him in the production of three small works to be performed by Juilliard students at the Juilliard School. This wonderful gift certainly cheered Tudor up. Another happy occurrence surprised him the following year when, in July 1970, Tudor was awarded an Honorary Doctorate of Literature from St. Andrew's University in Scotland. He was delighted, and in apparent gratitude bequeathed them a sizeable legacy.

In July, with the modest $10,000 from the N.E.A., Tudor rehearsed and mounted *Three New Pieces for Small Groups:* String Quartet No. 1, "The Kreutzer Sonata," by Leos Janacek (the dance was renamed *Sunflowers*); Quartet for Percussion, "Inconsequenza," by Geoffrey Gray (the dance was renamed *Cereus*); and *Le Canon* by Pachelbel (renamed *Continuo*).

These ballets were intended for performance in the repertoires of smaller ballet companies, which were increasing in number in the United States. All three pieces were performed by the Juilliard Dance Students in the Juilliard Theatre. This private showing, which allowed no critics to review, conformed to the terms of the N.E.A. grant to Antony Tudor. Besides, Tudor never liked what the critics said about him.

Sunflowers (May 27, 1971)

In a letter to Isabel Mirrow (Brown), former dancer with Ballet Theatre, November 12, 1985, Tudor offered some valuable pointers about *Sunflowers*. He explained:

> The ballet's name directly evokes sunflowers which always keep their heads turned toward the sun, and bloom in the heart of summer. The four ladies of

the ballet are friends of long acquaintanceship having known each other for almost as long as they can remember and every summer they seem to have been brought together at this particular bit of countryside, for here they regathered in what seems to be the corner of a hayfield with an old weather-beaten log fence separating it from the country lane in the background and down which the two men have strolled onto the scene. They are also old acquaintances and have wandered by wondering if the girls are still around. I refer to it as my Tschekov [*sic*] piece probably because of the Slavic musical score, but also because I felt an affinity with plays like the *Three Sisters* and *The Cherry Orchard* etc. with which you are surely acquainted. The steps contain within themselves the aspects of the four girls. The youngest, and flightiest, wears a dress reaching to just below the knee, the next is about $1\frac{1}{2}$" longer, the third still about $1\frac{1}{2}$ inches longer still, and the eldest and primmest of the four, Airi originally, has her skirt hem about the bottom of the calves. The period of the costumes is the period of the above plays, and the youngest one's dress could be very like that of the younger sister in *Pillar*. The sleeve lengths and necklines differ similarly from one another conforming with the skirt lengths. At the opening of the piece, the four are full of "Ennui," and are resolutely determined to be happy and carefree together. When the men appear on the scene the true natures of each of the women evidence themselves, and by the time the two men have been able to extricate themselves from this situation, the four women are left with their true natures showing and the piece ends with them alone and separate. There is a painting, famous, by an American artist of a lady in white lying in the centre of a green field. . . . I think he was Thos. Eakins [actually it was "Christina's World" by Andrew Wyeth] but I'm not sure . . . and this is rather what I had in mind for their last posture. She is wearing white for the white and off-white were customary for summer frocks at that time. Voiles, ninon, light cotton crepe and so on. . . . Also to be remembered that, as in all Tudor pieces, the torso is the most important part of the body, and the legs move in order that the body shall be carried where it needs to go, NOT because the legs are doing something that makes the body go along.

In the Labanotation score, Tudor coached the notator by suggesting that the girls in the ballet, *jeunes filles en fleur,* are carefree, friends for years, and are the ones seeking love. The boys feel nothing. According to Tudor, that is their drama.[57]

In the interview with Marilyn Hunt, August 22, 1986, Tudor recalled that "when I did *Sunflowers* first, it was at Juilliard, and I modeled the girls and their parts on the people that were there as students. It always becomes slightly tailor-made." He described one of the moments in *Sunflowers* that was particularly "talented." "When those girls start walking, they do exactly the same step, but each time it looks different because they're wearing their arms in different positions."[58] When Hunt asked how he went about fitting the music with the scenario, Tudor sighed:

> I'm uncanny, that's why. I never know how those things happen. I just stand there and wait for something to happen. If it happens, it happens. If it doesn't I wait a little longer, and then it does. . . . I had the music first, and I had girls that wanted to work in something which pushed them all together.[59]

The ballet begins with four young women, walking and doing common folk movements. The group reduces to three young women; they walk around each other in a circle with very open torsos. They seem to be rotating as the planets do. Four of them form different patterns with one in each corner; one comes center; they all circle and repeat early movement phrases. The violins nostalgically sound a touch of pain. The dancers travel in spinning turns like whirlwinds. One girl begins to move through the others; three join her; the square is curiously satisfying as a spatial design. Tudor uses the Wyeth pose of the woman on the floor reaching into the distance. From the beginning, the ballet unravels in patterns of square-dance forms and interweaving contra-dance shapes.

In the second movement, a girl performs a series of jumps with one leg in front, then jumps quickly lifting both legs underneath her; she moves with her arms high. On the right diagonal, two men enter with women partners, but seem to be independent of them, and exchange women; one couple leaves; one stays on stage with the lone figure of a woman behind watching them. They execute consecutive attitude turns facing away from the audience and toward their partners. The young woman who remains holds her face, covering it with her hand. The pair moves together doing turns with their legs high out to the side. As a couple, they seem to be opening up like flowers to one another. Then the first young woman moves in; two young women edge in towards the young man. The other couple enters. The young men lift their partners and move rapidly to the side. When one of the women is lifted, her torso flies back while she travels forward. The other couple stays alone on stage and dance a passionate *pas de deux* as the violins crescendo. All the dancers enter breaking up into groups, with one woman tagging along. When the young men leave, the four women repeat the opening pattern. In the third movement, there is a three-and-one spatial arrangement of the women when the two young men enter. They stand while the young women move around and near them. Then they circle the women and begin to do lifts; one couple ends the section down left while a single girl is upstage right, facing the left diagonal downstage. The men leave, and the movement quality of the women becomes increasingly frenetic and frustrated. At the conclusion they repeat the earlier Wyeth pose.

Camille Hardy remarked on the changeability of the youthful women's intense emotions. "This is an interplay of feelings that can fly from insecurity to boldness to despair in hardly more than a moment."[60] Hardy goes on to analyze Tudor's way with the women "sunflowers." "The choreographer has not given us a flirtatious contest with the men as soon-lost prizes. It is a reverie of youth's painful, ineffable discovery that nothing will ever be the same again. Tudor's bittersweet sense of the precarious is recaptured by the dancers at the end as all four depart, each in a different direction."[61]

Sunflowers had its professional premiere with the Omaha Ballet Society on May 13, 1972. It was performed with *Fandango* at University of California at Irvine (April 1974), the Milwaukee Ballet (1976), the Kansas City Ballet (1978), the North Carolina Dance Theatre at Spoleto (1978), the Fort Worth Ballet (1982), Finis Jhung Chamber Ballet U.S.A. (1985), the Louisville Ballet (1986), and the University of California Irvine Dance Ensemble (1988), among others.

Cereus (May 27, 1971)

Cereus was inspired by a student party that Tudor attended, where apparently he was shocked not only by the promiscuity but also by the cruelty of these young people when one person was ignored or left out of the group's activities. Many of Tudor's ballets hinge on the ways men and women discover partners in love. Several years earlier, in *Knight Errant,* Tudor focused on a courtier and his conquests of aristocratic women at court. With *Cereus,* the naked and casual sensuality of modern American youth is explored to an upbeat and jazzy score. Tudor discovered some very appropriate percussion music with strongly accentuated rhythms. He distinguished the ballet with high energy, quick gestures and moves of young people, with hints of nasty practical jokes and fights over one girl.

The program notes indicated that *"Cereus* takes its name from a kind of cactus whose night-blooming flower of luxurious perfume appears once a year, a metaphor for youth."

The ballet opens on a male dancer, his weight on his back leg, the body tilting back but ready to spring. Gradually three more men enter. They are all wearing sleeveless tank tops, almost bare chests. In the notes that Tudor wrote for the Labanotation score, he stipulated that "the opening male solo is very nervous under a casual façade. The men's dance is low to the ground. The movement should be performed as large as possible. The jumps are out rather than up. The style is somewhat abandoned rather than neatly classical. This is the generation directly following the *West Side Story* gangs."

In the second movement, big, strong, rhythmic music impels deep *pliés* and free-feeling moves across the stage, highly un-Tudorlike, and punctuated thrusts of movement. A woman enters, then another, then another, on pointe in strapless tops and short skirts. Tudor suggested in the score that in their partnering with the boys, the "girls are like tendrils of a vine that curl around a stem." Seven young people make modern jazz moves, and from the odd number it is assumed that one person will be left out. Three couples dance together while a single man watches, looking quizzical. He follows what they do. The spinning of girls going from one partner to another creates the image of separate planets twirling in their own orbits, bouncing off other bodies, and young people choosing mates with a chaotic and frenetic energy.

Once again, in the notation score, Tudor comments that "another image is that all the dancers are like solar bodies which slowly revolve around one another. Move through positions, do not hit them. Nothing is percussive but rather one smooth phrase." These seem like difficult footnotes for movement to a percussion score. We focus on one couple who at first seem playful but later slightly combative. From time to time, the lifts, when a man throws the woman around, verge on violence. While the women are conventionally feminine, seductive creatures, the solos for young men are almost wild and very sexual. Three couples rush in doing unison partnering with linear and sharp moves. Two men who are lying down on diagonals are joined by a woman. At the end of the ballet, three couples remain together while one abandoned woman faces them. She gets pushed out of the group to discover the fierce games one encounters in courtship.

As in many of Tudor ballets, *Cereus* does not emphasize the heroic nature of people. Rather it portrays the poignant, tentative and occasionally brutal qualities of particular individuals; Tudor carefully delineates the lines of motive and behavior that go through their lives. We are always acutely aware of the human element.

When *Cereus* opened in Philadelphia, dance critic Judith Hansen found the ballet a subtle work but puzzling. The unconventional percussion score reminded the reviewer of a Balinese gamelan, "while the movements were detached and marionette-like at times but also smooth and natural and somewhat sensual at other times."[62] One critic called it "an interesting failure,"[63] while another critic enjoyed the music with its "jazzy underline for rubbery pelvises and angular poses."[64] Recently, Gus Solomons discussed the Alberta Ballet's 1990 production. "*Cereus* begins with a section of strong dancing for five men with unexpected directional changes and difficult combinations that distinguish Tudor's choreography. . . . But with the entrance of four women the piece becomes uncomfortably symmetrical."[65]

223

According to Jennifer Dunning, who reviewed *Cereus'* performance by the Bernhard Ballet of Connecticut at the Riverside Church in New York, "the ballet recalls the swinging ponytails of Jerry Robbins's jazz ballets. Changing partners inevitably isolates one dancer or another, if only for an instant, and along with childish games, dark moments of stillness are interspersed in the men's jazz lunges and cool swivels, the women's delicate struts and well-executed, witty off-kilter lifts."[66]

Cereus' first professional performance was by the Pennsylvania Ballet (1972), and subsequently by the Bernhard Ballet in Connecticut (1979), Juilliard (1986), the Princeton Ballet (1987), and the Alberta Ballet in (1990).

Continuo (May 27, 1971)

Tudor was never the one to shy away from a piece of music that might seem daunting to a less secure choreographer. His *La Gloire* (1952) to several Beethoven overtures, including the *Leonora,* may not have succeeded as a ballet, but he certainly knew how to use the majesterial music. In the case of *Continuo,* he discovered the unique lyricism of the Pachelbel baroque score during a period when many music lovers were singing the tune as they walked along the street. He had to create a dance piece that would enhance the music and heighten its already-well-known melody.[67]

Some notes from the Labanotation score written by Tudor mentioned that *Continuo* is a non-programmatic "classical ballet. . . . It is one long lyrical out-pouring. The movement flow never stops. The 'steps' should be linked with freedom and abandon." This quality of linkage characterizes Tudor's movement style and is often the most prominent feature noted by critics.

The ballet opens with a couple entering stage left. The girl is lifted, doing careful, fluid attitudes, *sautés,* and turns. She seems to be constantly aloft. Her partner promenades in an *arabesque plié;* she follows him. The second couple enters and all four people dance together, often on a diagonal, then in a circle, with swift changes of weight and lifts in *arabesque.* The prominent movement image during the canon music is a breathtaking swoop lift with a fish dive image (described below by Kisselgoff) as well as lifts in *passé.* Tudor's movement combines with the music in a delicate and understated way. One couple replaces the first with a particular movement phrase but does not displace them. The young women are pulled into the air and seen to fly. The first couple leaves and the third couple enters, leaving four people on stage. Once again they work on a diagonal line with *chassés* and *piqué passés;* a solo woman does *arabesque piqué,* jump front. One man joins her; they work in the center. A man jumps front into *arabesque,* then does an *arabesque* turn while his back leg comes into *plié* in a turned-in spin with one foot on the ankle of the standing leg. Three men on the diagonal rush from right to left to meet their partners. Finally all six people dance together, the center couple working against the outer two. In the last phrase, the men are on the floor on one knee, the other leg out in *arabesque.* At this moment, the women hold the men's arms out to the side while two of the couples face diagonal corners.

Several characteristic movements of Tudor ballets highlight *Continuo:* the transference of weight on pointe from one foot to the other in fifth position while the upper body turns and swirls, and the movement the men execute as they swing their partners, who seem to fly with fluttering kicks behind them. In the meantime the men do very stylized, Renaissance steps, stepping diagonally with the foot moving to the front. Tudor also liked the angle of turns with one leg in front and *pointe tendu* front. In addition there are many deep *demi-pliés* with *relevés,* giving a sense of soft downs and ups.

Anna Kisselgoff attended a Brooklyn College showing of *Continuo* by the Joffrey II dancers. She wrote:

> Some of the most beautiful choreography in existence. . . . The continuum implied in the title is very much a formal theme in a ballet whose dancers never stop in the flow of movement that sweeps across the stage so naturally. . . . It is all the more startling then to see the filigree tones interrupted by an inventive flying leap in which a boy yanks a girl forward by the arms, her legs scissoring behind her, or the unexpected moment when a repeated pattern is broken by a girl falling out of line.[68]

All three N.E.A. ballets (*Sunflowers, Cereus,* and *Continuo*) provide rich sources for small companies and enable them to mount a Tudor work without the cost of expensive sets and intricate costumes.

The first professional performance of *Continuo* took place in New Haven, Connecticut, with the Syracuse Ballet Theatre on February 15, 1976. It had not been performed since 1971. *Continuo* was performed by Joffrey II (1977), Metropolitan Opera Ballet (1981), and Joffrey II (1981). The first French performance was at the Paris-Opéra Comique, "Hommage à Antony Tudor" (1985). It was also staged by Boston Ballet II (1986), Ballet van Vlanderen (1987), and the Alberta Ballet (1990).

That summer (July 1971), American Ballet Theatre revived *Romeo and Juliet,* which had not been seen since 1956. Arlene Croce atypically lauded Tudor's 1943 work: "The ballet is not Shakespeare's tragedy, but a shimmering vision of the Renaissance, weighted by one of the most splendid architectural sets of modern times. Tudor's choreography curls around it like an insinuating growth."[69] Croce made the curious point that Tudor reveals more about the 1940s, when he choreographed it, than the Renaissance: "Its deliberate aestheticism, its almost too exquisite touches are as absorbing as a brilliant nightmare. . . ."[70]

The same year that *Sunflowers, Cereus,* and *Continuo* were created, Tudor retired from Juilliard after twenty long and fruitful years. Later on he confided that twenty years of teaching at Juilliard was quite enough and that he hoped for a pension that would make his elder years easier. His work at Juilliard provided an excellent model for the years to come. He created a fine professional ballet program of strong and creative dancers who have made major contributions to teaching and choreography throughout the world.

The following year, on the invitation of his Ballet Theatre colleague and friend Eugene Loring, Tudor joined the faculty part-time at the University of California at Irvine, where he worked periodically and seriously until the last class he taught in 1986. When Tudor retired from Juilliard, he had immediately sought another educational tie, though his relationship to Irvine was casual, teaching young dancers and choreographing at a leisurely pace. His relationship to the West Coast and Los Angeles grew as he and his friend Hugh Laing began to spend the cold New York winter months in sunny climes near the sea at Laguna Beach, which is close to U.C.-Irvine. His influence at Irvine should not be underestimated; he not only lectured and gave classes there, but he also restaged some of his chamber ballets. In "A Loving Memoir," Olga Maynard recalled that Tudor's presence at Irvine "was a daring experiment," a most successful one, and that "Tudor lent éclat to the UC Irvine campus."[71]

CHAPTER NINE

Honors at Last

I only know that if we don't grow constantly we die automatically.
While we stick to the familiar we are retrogressing and not going ahead
the way we should.[1]

—Antony Tudor

During his last years, Tudor traveled in America and Europe to re-create many of his ballets. He visited Hugh Laing's home in Barbados in 1972, and England and Düsseldorf for reconstructions of his ballets in 1973. He was also awarded the precious Carina Ari Gold Medal in 1973.[2]

After much dedicated work in New York, he received the coveted *Dance Magazine* award in 1974. The citation read: "To Antony Tudor—a monumental force in dance; one of its great original geniuses. He, no less than his predecessor, Michel Fokine, and his compeer, George Balanchine, has defined the nature of twentieth-century dance."

In his inimitable style of subtle mockery, Tudor looked at the other choreographers who also were receiving the *Dance Magazine* Award and said, "These two young men, Gerald Arpino and Maurice Béjart—prolific, producing marvelous pieces one after another for these last few years. I haven't done anything for twenty-five years. And then I knew! They gave me the Award for not doing anything."

How seriously should one take Tudor's modesty on this historic occasion? One might doubt that Tudor wanted to be compared to Arpino and Béjart, regardless of their output. Tudor then offered thanks to his old friend Hugh Laing, "an extremely harsh critic who really was very rude sometimes, extremely rude and unforgivable. . . . But nevertheless he put me on the straight and narrow path."[3]

One of the most significant events of his life occurred in 1974, when Lucia Chase and Oliver Smith brought Tudor back into the fold. He was appointed associate artistic director of the American Ballet Theatre, sharing his appointment with Nora Kaye, while Chase and Smith stayed on to deal with other problems of the company. Nora Kaye had already had a brief involvement as an assistant director in 1964. Tudor's ballets had always been the backbone of Ballet Theatre, and for many years he sustained crucial ties to the company. He felt, however, that he had lost his voice in its creative development. In a sense, this was a final vindication and a long-hoped-for honor. Tudor remained at Ballet Theatre as artistic director for four years.

During the stimulating year of 1974, he traveled to Sweden for the televising of *Dark Elegies,* although he did not stay long enough to see it through to the end. As a result, there were serious complaints by the Swedish dancers about the final version.

A surge of new vitality and purpose occurred after Tudor assumed his new position; he restaged *Shadowplay* for Baryshnikov and Gelsey Kirkland and created one of the most poetic ballets of his career, *The Leaves Are Fading.*

The Leaves Are Fading (July 17, 1975)

For the *Leaves Are Fading* score, Tudor chose Dvorak's little-known "Cypresses" together with the String Quintet, Op. 77, the String Quartet, Opus 80 and part of his "Terzetto."[4]

Critics accused Tudor of a seven-year choreographic silence before the making of *Leaves.* It is important to remember that although Tudor, now sixty-six years of age, had not made a big ballet since *Knight Errant* (1968) for the touring company of the Royal Ballet and *The Divine Horsemen* (1969) for the Australian Ballet, he did choreograph several "little ballets" for Juilliard in 1971. His loving and wistful *Continuo,* his strange and disturbing *Cereus,* and his compelling *Sunflowers* were by no means negligible artistic projects.

In an interview with Marilyn Hunt, Tudor spoke about the fact that *Leaves Are Fading* is inspired to some extent by *Sunflowers.* Hunt made the comment that

> there seemed to be a progression one could make from the summer of *Sunflowers* to the fall in *The Leaves Are Fading*—both in the sense of seasons and human life. . . . It's as if the people in *The Leaves Are Fading* are remembering episodes, not the same ones, of course, but some similar feeling of remembering summertime.

Tudor answered Hunt:

> In *Leaves Are Fading,* all the couples who dance together are in love, one way and another. In the other piece, the girls are in love, but the boys aren't. It makes a big difference.[5]

The Leaves Are Fading re-established Tudor's pre-eminent position as a great choreographer at American Ballet Theatre. He often remarked that he did not want to repeat himself, and indeed this ballet stands out as unique in his repertoire—with its plotless, rhapsodic expression of couples in love. "Tudor had indeed erased all signs of 'personal' technique so that the viewer sees only the movement, the pure, seemingly spontaneous action. This is the transcendence of form Tudor himself described as 'empty'—to use a Zen expression—empty of ego, he seems to mean."[6]

During the rehearsal period for *Leaves,* Tudor began to work with Gelsey Kirkland, an exceptional talent who had been trained by Balanchine and who was now working with American Ballet Theatre. He knew that she had a refined sensibility and a strong personal integrity. A short time later, Gelsey was dancing in three of his ballets: *Jardin aux Lilas, Shadowplay,* and *The Leaves Are Fading.* In her book *Dancing on My Grave,* Kirkland refers to Tudor's methodology in the studio: "Each Tudor dance that I performed was both a fantastic puzzle and a social portrait composed of intricate ballet images and mimetic gestures. Each detail held a possibility for revelation." At one point during rehearsals, Kirkland stated, Tudor decried the way another dancer was overacting, "tearing a passion to tatters."[7] Not long afterwards Tudor wrote a similar note of criticism to Kirkland and reminded her that the next time she performed *Swan Lake* she must not "fall into personal melodrama."[8] She appreciated both Tudor's "flashes of caustic wit" and his profound reverence for the theater. They had something important in common. Kirkland dedicated her second book, *The Shape of Love,* to the memory of both Tudor and Frederick Ashton.

With a relatively large cast of fifteen dancers, *The Leaves Are Fading* revolves around a series of passionate *pas de deux.* It begins and ends with a single young woman in a long green summer dress drifting on stage, her hands behind her back. She looks around, up and down stage, seeming to gaze into her past and perhaps our past. She slowly crosses the scene into shimmering early autumn shadows. Gradually eight women dressed in simple, leaflike chiffon dresses fill the stage, dancing in patterns of four and three until two men fly into the group to initiate leaps and lifts. The men organize into a circle doing slow *grands jetés,* their legs spread far apart at the height of their jumps. Typically Tudor borrows from an East European folk-dance vocabulary suggested by the music. These dances introduce a weightiness and staunchness or solidity to the partnering work.

The first couple emphasizes quick and exacting allegro steps. She is lifted in airy, carefree passes, and spins with low *piqué arabesque* turns. The two figures spread out the length and width of the stage. One particular *jeté* lift transports her far backwards as she faces the audience. When listening closely, the viewer notices that Tudor often carries the dance phrase beyond the musical phrase.

As the first couple leaves, young men enter in a series; first two, then three, etc. Wending their way in jumps with both knees bent, jumps with one leg to the side, hitch kicks in back attitude, they seem to be doing a Slavic folk dance. When six women return, they execute very formalized technical steps with a rather Baroque flavor. They divide into couples, and then the women dance for the men, who retreat to corners of the stage.

The second couple, originally played by Gelsey Kirkland and Jonas Kage, offers a totally different energy and romantic mood to their dancing. Their love seems to be a "sweet love remembered." His stolid and weighted steps contrast with her willowy and delicately fluid movements. He whisks her into *arabesque* and drop-lifts her to swing around him. Once more he takes her aloft and turns her high with one leg to the side. She tantalizingly dances for him with her arms over her head, and he hurries her into a *pirouette* lift into attitude, recalling the backwards jumping lift of the first couple. She finishes sitting on his knee. Their duet seems to symbolize a yearning for the past, or the future; what might be imagined rather than what is. Several couples re-enter, an interjection of society that surrounds and breaks the singular intimacy of lovers.

The third duet reveals a more sensitive and serene amour, less ecstatic and more tender. This woman has an elegant and smooth confidence about her. The couple move as one; her clear connection to him emphasizes his powerful and stalwart qualities. Her arms reach for him. He lifts her with one leg in front of his shoulder, the other behind. She pushes down on his hand, her leg behind, and effortlessly turns.

The last couple dances, nervous and fleeting with a burgeoning passion. Playful, excited, and youthful, they quickly achieve perfectly matching lines that are enhanced by their spontaneity and clarity. The second couple return in an outburst of pure windswept movement. She is supported, falls backward with complete trust, is swept into a lift that flies joyously skyward, like a wafting breeze carries a leaf. He lifts and parades her high, and like a ship's prow they rush back and forth across the width of the stage.

Once again all the couples return. There is the feeling of a garden or a park as people stroll in and out with a sense of leisure and love. Once again the single woman enters, stage left this time, carrying a red nosegay; we are left with the sense of having accomplished a journey. She crosses the stage and leaves on the side of her first steps.

In her article "Sweet Love Remembered," Arlene Croce commented that:

> *The Leaves Are Fading* is much more a return to the enduring strengths of the classic Tudor ballets. What it requires of its dancers—subtle musicality, sus-

tained line, clarity of nuance within a phrase, are also required by *Jardin aux Lilas, Pillar of Fire,* and *Romeo and Juliet.* The difference is that in the new work these qualities are rendered for their own interest as pure dance qualities; they're not intended to support dramatic meanings.[9]

John Gruen quoted Tudor about *Leaves:*

Well, yes. It's a piece without any story at all. Of course, if I see a human body on the stage, I don't see it as an abstraction. I see it as a body. So I would not call this new work of mine "abstract." Rather, I would call it empty.[10]

After the initial success of *Leaves,* it became an important subject of conversation that Tudor had created a ballet without a serious story. After all, Balanchine's fame as a choreographer rested on the notion that ballet no longer needed a dramatic scenario, that the movement carried the message. But when interviewed by Zita Allen, Tudor rejected the terminology by saying that "I can't choreograph an 'abstract' or 'empty ballet.'" But neither did Tudor like getting entangled in the form-versus-content wrangle. "I don't think you should be able to categorize them and separate one from the other in my kind of work." As for the label "modern realist": How would he distinguish between his starkly poignant ballets and the champions of romantic illusion like *Swan Lake?* he was asked. Tudor quickly responded, "I don't use feathers."[11]

The Leaves Are Fading continues to be performed by Ballet Theatre. It was re-created in London by the Royal Ballet (1975), in Berlin by the Deutche Oper (1984), in Paris by the Paris Opéra Ballet (1986), and in Leningrad by the Kirov Ballet (1992).

Tudor was invited again to Israel in the fall of 1976 to recreate *Dark Elegies* (November 11) for the Bat Dor Company. During the same year, he received the Brandeis University Creative Arts Award. Throughout his life Tudor felt slighted and ignored by the people who pronounce fame and bestow fortune upon their subjects. Fortunately, Tudor finally began to acquire the recognition he deserved.

In 1978, Tudor turned seventy years old. Unfairly, this year proved to be one of the most disheartening and difficult of his existence. Though he was feeling tired and worn out, he agreed to choreograph a rather bizarre work, *Tiller in the Fields.*

Fernau Hall alleged that "just before *Tiller* in 1978, they told him he was not to become Artistic Director of American Ballet Theatre; rather it was to be Baryshnikov."[12] It was not until 1980 that Baryshnikov became artistic director, however. Tudor stepped down as associate director of Ballet Theatre on September 16, 1978, and he then had a heart attack, which followed quickly on the heels of *Tiller's* opening. The rehearsals and the worrying contributed to his deteriorating health.

Tiller in the Fields (December 13, 1978)

Zita Allen asked Tudor how he came to make the ballet *Tiller in the Fields.* Tudor sarcastically replied: "The inspiration for *Tiller* comes very simply. At a company meeting someone asked Mr. Baryshnikov what he would like for the future. He said he would like a new classical ballet made for him to dance—something like *The Leaves Are Fading."* Tudor took time to think about this request and said, "I'll go look for some more Dvorak music and make the ballet for him, maybe to satisfy him—a classical ballet in that style, in a similar manner. Only he left the company to dance at New York City Ballet. But I continued to make it anyway. This is my childish piece. Why? Because I'm getting into my second childhood."[13] Tudor was asked, "Is *Tiller* about Zen?" and he retorted, "No, *Tiller* is about sex."

Tiller in the Fields continued Tudor's previous exploration of themes that are served by the wondrous, occasionally menacing, backdrop of nature. In *Shadowplay,* the Boy with Matted Hair struggled with the Arboreals and the Celestials in a world that is threatening, but ultimately comforting and accepting. *The Leaves Are Fading* embodies the romantic theme of love heightened by the concept of autumn leaves, fading and falling and swirling in the wind. In *Tiller,* nature and the passage of time surely and quietly take their course.

Like *The Leaves Are Fading, Tiller in the Fields* was set to excerpts of Dvorak compositions and has a similar pastoral setting. The time is between spring and summer. The landscape is distinctly European, with a cathedral rising above the fields of the painted backdrop.

The Tiller, a young peasant boy played by Patrick Bissell, opens the ballet with a spin and a fall to the ground landing in a languid posture that becomes his motif. It seems that he loves the sunshine far more than either work or women. As the village women cross the stage, he dances with one and melds into the ensemble of six couples with mazurka-like pushes of the heel into the floor and patterns of circles that fly out into space; what Deborah Jowitt calls "fresh-air classicism."[14]

For the scherzo from Dvorak's Sixth Symphony, the opening *furiant* is quite literally a peasant dance for four high-flying men. The soldiers in *Echoing* also pounded the heavy, peasant-like figures, but they bore ill tidings. It was Tudor's former ballet *Sunflowers,* Kisselgoff suggested, that was the paradigm for these folk forms. But Tudor had been using folk forms for

Tiller in the Fields, *1978. Gelsey Kirkland and Patrick Bissell.* (From the Performing Arts Research Library)

years. In *Sunflowers,* Tudor "delicately suggests girlish friendships disturbed
by the intrusion of two boys, with these stirrings of adolescent love and the
inevitable changes in the course of life." In the same way, *Tiller* introduces
the outside person, this time a woman. Very dramatically, a different kind of
figure intrudes into this bucolic setting.

Kisselgoff describes Kirkland's role: "It is a gypsy girl whose disturbing
presence is rendered marvelously and insistently by Gelsey in her expressions and quality of movement."[15] With one arm slung thoughtfully over
her head, she immediately attracts the male with her soft dancing. Like
heroines of romantic ballet, she appears only when his friends have left the
scene and disappears when they return, unnoticed. He wonders whether she
is just an ideal figure that he has imagined. She will become the agent of
experience who destroys the earlier innocence of the hero.

Both characters are, in a sense, prototypes of romantic literature. The
gypsy seductress has a buoyant-as-breath quality in her leaps, a breezy flail
to her arms, and a childishly pouty quality to her sexual presence. Another of
the ballet's signature moments focuses on Kirkland lying flat and being
carried happily about like a rag doll, or more to the point, a sheaf of wheat.
Meeting after meeting is suggested, and the *pas de deux* culminates in one-

233

handed lifts, through *pirouettes* in which Bissell's mere touch sends Kirkland into spinning turns and include a brief variation with Bissell showing off, before the spirit of the watching girl-woman seems to capture him for another run to the woods. During this duet, Tudor quotes his *Pillar* dialectic of "lovers in innocence" and "lovers in experience" where sexual love is pitted against romantic love. He accomplishes this reference by counterpointing the ecstatic *pas de deux* with two couples in the rear who seem to represent a quieter, more sustained love relationship.

The small gestures are telling, as usual with Tudor. The Tiller runs off after the gypsy with his jacket as if to protect her from the coming cold winds. For the conclusion, Tudor has Kirkland come out wearing his jacket and looking very vulnerable and very rounded. Bissell's reaction is to throw his hand to his forehead in horror. Kisselgoff asked, "Is this Peyton Place in a Slavic town?" After Bissell has recovered from his shock, he runs happily offstage and into the future with Kirkland.[16] Deborah Jowitt summed it up: "The planting of a seed, new life, fresh growth, the tiller of the soil. Kirkland is the young, rural fertility goddess."[17]

Tudor was interested in provocation, in setting people up on their toes, and in causing a ruckus. The *coup de théâtre* in *Tiller* turned out to be the gypsy girl's pregnancy, a device that ultimately dismayed and perplexed the viewer as, with this event, the whole tone of the ballet was transformed. The "mixing of genres" displeased New York audiences. According to Kisselgoff, "none of this works. . . . Judged on the choreography alone, the ballet is brilliant, even wholly fresh in many combinations of steps and shapes." Kisselgoff allowed that anything on stage is permissible if it works. But there is a sharp change of tone between the stylization of the overall choreography and the naturalism of the final scene.[18]

Tiller was interpreted in different ways, as allegory or metaphor. In the *Press Journal*, Dorothy Thom pronounced Gelsey Kirkland "the goddess Ceres who comes to bestow her blessing on the young farmer and his crops. But what about her crop?"[19]

One of the more illuminating columns concerning *Tiller* was written by Jack Anderson: "Ballet Theatre to Close with *Tiller in the Fields.*" Anderson speaks of the oddities in *Tiller.*

> It starts in fantasy and ends with the facts of life, that the source of this strange tale parallels the libretto of "The Diary of One Who Vanished," the song cycle for alto, tenor, small chorus and piano that Leos Janacek composed in 1917 after reading a set of anonymous poems in a Czech newspaper.

The poems told of a peasant's love for a gypsy, and Janacek considered them notable examples of twentieth-century folk poetry. Tudor, an aficionado of Slavic music, used Janacek's music for his ballet *Sunflowers* in 1971. When Anderson asked Tudor if he had heard of Janacek's "The Diary of One Who Vanished," he quickly replied: "Oh yes. I know it quite well, I wanted dearly to have used it for a ballet, and for a long time I thought I was going to use it for this ballet." Tudor explained that he did not refer to it because Janacek's approach to the story did not coincide with his own, which demanded a youthful, "second childhood" approach. Anderson alleged that *Tiller* remained deliberately naïve and childlike until the ending, which jarred. Tudor asserted that "that scene can be legitimately played for embarrassed laughter—for that's the kind of laughter it really is, you know. It's the embarrassed giggling that can come when people feel awkward or uncertain about how to react to something."[20]

Though Tudor chose the exuberant and robust music of Dvorak, and not the introspective music of Janacek, the story of a girl getting pregnant remains. In a way, in this last ballet, he harks back to the *Pillar* theme where a woman is lustful and seduces an attractive man. In both scripts, the pregnancy results from the actions of women committed to their own needs, and Tudor does not present them as innocent beings who are not responsible for their actions. Both Tudor's heros and heroines have choices and make decisions knowingly. One suspects that Tudor might have reworked this ballet if given the time and encouragement. The ballet has not been performed since 1979.

Tudor recuperated from his heart attack in Laguna Beach, where it was quiet and less stressful than New York. Laing provided the necessary care for healing him. In a letter to Martha Hill from Laguna Beach on January 29, 1979, Tudor was joyful that his "old age" pension had just begun, and "that little heart of mine rebelled at such a marvelous time."

After this crisis, Tudor never accepted another choreographic assignment. He gave the job of restaging his ballets to Laing, Sallie Wilson, Airi Hynninen, and others. These devoted friends took the responsibility of keeping his *œuvre* alive and well. Tudor continued to consult, advise, and oversee the last rehearsals when possible. In the 1980s, Tudor's most popular ballets were frequently performed by American Ballet Theatre, as well as by large and small ballet companies around the world. His reputation finally achieved the weight of a great creator.

When Marie Rambert died June 12, 1982, Tudor was invited on the occasion of her death to write "Getting to Know Rambert" for *Ballet Review*. His words reverberate with conflicting but sympathetic tones:

> One cannot speak of Rambert without the use of excessive terms, for her superabundance of energy and her all-consuming enthusiasm were devouring,

and she insisted that they be shared by everybody around her. Nor could she be contradicted. . . . There was so much to learn from her—from this brilliant little woman who devoted her talent to the fostering of her aims through her students and who gave unstintingly of herself in the process.[21]

A few years earlier, Tudor had written "Mim" a letter in August 1974 acknowledging how "fortunate to find myself (and Frederick Ashton too) in your magnetic field. . . . I inherited staying power from you. Audacity? Patience? Tolerance? NO! However, more probably a being made more sensitive to our instincts and the ability to see and smell out the right."[22]

Tudor also survived a love-hate relationship with the other powerful woman in his life, Lucia Chase. When she became very ill and bedridden, Tudor dutifully visited her and sent notes, books, and well-wishes. In the end his essential loyalty prevailed, but when they were younger, he resented her dominant position. She died in 1986.

As a measure of Tudor's increasing popularity, the Paris Opera asked him to put together an evening of his works in February 1985. Tudor, who loved Paris and adored French food, was finally invited by Rudolf Nureyev to produce an evening of his ballets at the old and beautiful Salle Favart of the Opéra Comique. Unable to withstand the strenuous schedule of rehearsals, he sent Sallie Wilson to restage *Jardin aux Lilas, Continuo, Shadowplay,* and *Dark Elegies.* During the next season, they added *The Leaves Are Fading* to the Tudor repertoire in Paris.

From this point on, Tudor began to receive a number of awards that recognized and celebrated his contributions to the ballet world. One might conjecture that, by now, Tudor's students had reached a time in their lives when they had enough power and fame to repay their teacher and bestow honors on him. In London, he received a certificate honoring him from the Royal Academy of Dancing for his ballets *Dark Elegies* and *Shadowplay.* These were the people who had originally certified Tudor to teach beginning levels of ballet and character dancing. On January 18, 1986, Tudor received the Dance U.S.A. Award, and on April 28, the valued Capezio Award. Eliot Feld, Paul Taylor, Agnes de Mille, Jerome Robbins, and William Schuman all toasted him with wonderful accolades.

Tudor had been living the life of a saintly guru. He spent hours working on his correspondence, especially with former students and artists with whom he felt a special rapport. He often dined with his old friend Hugh Laing.[23] He had mellowed, although the sarcasm had not disappeared from his bantering style. He continued to live at the First Zen Institute, where he led prayers and oversaw many duties. In a disciplined way he separated the religious area of his life from his work, a characteristic that puzzled both friends and critics.

Tudor died of a heart attack on Easter Sunday, April 19, 1987. Despite complaints of fatigue, he had been hard at work on rehearsals for a revival of his *Pillar of Fire*. It opened at the Metropolitan Opera House just one week after his death. At a memorial service for Tudor, Mikhail Baryshnikov, who was then the artistic director of American Ballet Theatre, spoke of Tudor's legacy. He reminded the audience that Tudor's ballets were made in the studio and that the young dancers who had been working with him had had a unique opportunity: "They took something away with them that can never be replaced, a sense of dignity and beauty and the importance of realizing our potential capacity to fuse heart and mind and body and music in the ways no other choreographer has."[24]

The last two prizes that Tudor received in 1986, the Handel Medallion of New York and the Kennedy Center Honors from the President of the United States, gratified him. Thus, in the year before his death, Tudor quietly savored a certain success that soothed the lingering pain of not having realized his potential as one of the greatest artists of the twentieth century.

CHAPTER TEN

Epilogue: The Heart of the Matter

*I knew I had to make my mark in the world; I knew I had something to
offer and I knew if I became a choreographer, it would be the easiest route
to become famous.*

—Tudor in conversation

*W*hen Tudor made this outrageous statement in 1986, there was a
twinkle in his eye. Choreography for Tudor had not been an easy way
to the ephemeral and tepid "fame" he was experiencing then. The combination of arrogance and self-mockery in this quote is typical of the man in his
later years. Perhaps he could not say straight out what his career proclaimed:
that he loved dance passionately, that he had quickly recognized his own
gifts as a maker of ballets when still a raw youngster at the Ballet Rambert,
that he had refused to compromise his aesthetic vision, even at the cost of the
glittering "fame" the young Tudor had undoubtedly dreamed of gaining. As
for the rest, Antony Tudor had "something" to offer the world, that much is
certain, and he did make his mark, one that is destined to last in the world of
dance.

The question of why, despite his genius, he did not achieve the popularity and success of his prodigious contemporaries remains unanswered.
Central to the issue is sheer output. Tudor did not produce the hundred
works that kept Ashton's and Balanchine's name before the public. Some
dancers and critics have theorized that Tudor was unable to reconcile his
perfectionist tendencies with his creative ideas. The problem with that explanation is that he was productive in those early years in London and, even
later, in the 'forties at Ballet Theatre. Another conjecture postulates that
Tudor failed to produce a large body of work because he was unable to
transcend his self-doubt. P. W. Manchester offered another idea: that Tudor
was thwarted after his initial fecundity for Ballet Rambert on the small stage
at the Mercury Theatre, because he found it overwhelming to work with
large casts of dancers and a larger stage, once he was engaged by Ballet
Theatre in New York. She remembered meeting him on the street and
asking when one might expect another ballet. "He answered more or less in

these words. 'I have to have something to say, and for years I haven't had anything I wanted to say.'"[1]

It would be well to remember that Tudor had practical handicaps as well as psychological ones. After leaving London, he never again had his own company, which meant he seldom had enough rehearsal time for his admittedly slow method of composition. Unions demand wages and sensible working hours! Exactly why Tudor did not form his own company in the United States is another puzzle. Possibly he did not know how to play the political and financial games needed to win over the money people who funded dance in New York. One might guess he was always a bit too prickly and difficult, perhaps too proud as well, to be good at raising money or charming millionaires. In the end, unraveling the mystery of Tudor's relatively low productivity may be a futile quest, but there is no doubt that all these factors throw light on the enigma.

Complicating the problem of evaluating Tudor's place among his peers is that he himself too often dismissed or criticized his work, refusing to keep some of his ballets alive, sometimes not wanting to discuss their value. He was also uninterested in preserving records, letters, and documents. His refusal to save gifts, his contempt for sentiment, his sense that nothing lasts: these aspects of his personality worked against him and hastened the extinction of some of his ballets. Was it that he truly did not rate them highly? Or did he believe that works of importance would somehow endure without his help?[2]

Tudor's dispirited reticence about his ballets was related to his acute sensitivity to criticism. All artists are the recipients of both fair and unfair critiques. Tudor, like most, had his share of both. But unlike thicker-skinned souls, Tudor suffered, perhaps unduly, when the inevitable imperfections were pointed out. Clive Barnes, a firm advocate of Tudor's ballets, defended him by saying, "even his so-called failures had distinction. Have any American classic choreographers, apart from Balanchine and Robbins, ever produced a brace of works as good as *La Gloire* and *Offenbach in the Underworld?* These are the kind of 'failures' on which lesser people have based careers."[3]

During the late 'fifties and early 'sixties a consuming debate filled conversations about the New York City Ballet and Ballet Theatre, especially in studio dressing rooms. It centered on comparisons between Balanchine and Tudor, and dancers and the dance world chose up sides. At the time, most came down squarely for Balanchine, whose ballets ignored or defied narrative themes while exploring a heightened, more athletic dance language. Balanchine's leggy, Olympian beauties bounded through articulated movements in luminous spaces, their Apollonian grandeur fascinating audiences and dancers alike. The New York City Ballet and Balanchine were the beloved of the media and money people, while Tudor's dancers resented the Russian choreographer's monolithic importance. Balanchine and the New York City Ballet had won the debate. For the next two decades, Tudor was relegated to the fringes: Juilliard, the Met, and Europe.

Nevertheless, Tudor had his supporters, but it took time for them to speak their minds. In 1972, Harold Schonberg, music critic and writer on the arts, defended Tudor while attacking Balanchine. Schonberg criticized Balanchine's plotless ballets for their "severe patterns . . . where humanity is lost," adding that "there is beginning to be a greater appreciation of Antony Tudor, who has not created a significant ballet for Ballet Theatre since *Undertow* in 1945. . . . Balanchine and Tudor are polar. Where Balanchine works in abstract patterns, Tudor is the exponent of the psychological ballet, the ballet with a story in which emotions are probed and motivations examined." Schonberg believed in Tudor's sensitivity to inward feelings. "Never sentimental, never a mere entertainer, never merely fashionable, he accomplished his aims with a taste and a Proustian feeling for character."[4]

Even at the pinnacle of Balanchine's success, when Tudor's ballets were more often performed and appreciated abroad than in America, cognoscenti like Schonberg recognized Tudor's genius. But in the chronically underfunded ballet world there was no room for two "temples of dance." Lincoln Kirstein and George Balanchine had created an empire of the School of American Ballet and the New York City Ballet with a flurry of promotion and systematic publicity. Their pre-eminence was confirmed by the fact that the New York City Ballet Company not only inherited the New York State Theater at Lincoln Center but also acquired valuable studio space needed by the Juilliard Dance Division. Antony Tudor, Martha Graham, and José Limón, all professors of dance at Juilliard, had hoped for a dance theater in Lincoln Center, a new, well-equipped space that would cater to all dance forms. It was not to be.

A curious article, "Good Guys *vs.* Bad Guys at Lincoln Center," by Douglas Turnbaugh, publicly mourned the situation, unfortunately when it was already too late to do any good:

> It looks like the ground has been sold from under the feet of the Juilliard Dance Department. This tenderly nurtured growth is probably going to be torn out by the roots . . . the cultural life of New York City is enfeebled and diminished by such perverse exclusions from our Performing Arts Center.[5]

We may question with other critics if Tudor was in tune with the aesthetic values that opened up, tripped up, and shook up the 'sixties and 'seventies generation of dancers and choreographers in America. Could he compete with colleagues who created *Movements for Piano and Orchestra*

(Balanchine, 1963), *Hair* (Julie Arenal, 1968), *Oh, Calcutta* (Margo Sappington, 1969), and *Rainforest* (Merce Cunningham, 1968)? How distant was Tudor's movement style from the artistic stream of the 'sixties?

It was both far and near. Even though major modern dancers such as Martha Graham and José Limón continued to make narrative works into the 'sixties, the vital signs of the "movement for movement's sake" were everywhere apparent. In those years, Tudor's *Echoing of Trumpets* was a major triumph in Sweden, and *Shadowplay* and *Knight Errant* had moderate successes in England, while *The Divine Horsemen* failed miserably in Australia. Even worse from Tudor's standpoint, only two of these interesting and quite innovative narrative scenarios made it to the major venues in the United States, and then only with very few performances.

These ballets were certainly a far cry from the pedestrian dances, happenings, and ritual occasions that monopolized American theater then. Except for occasional revivals by Ballet Theatre of *Jardin aux Lilas, Dark Elegies, Gala Performance,* and *Pillar of Fire,* Tudor's work virtually disappeared from the big stage and the Big Apple scene during the 'sixties.

An artistic current in the 'fifties that helped shape the next two decades of performance is pertinent. After World War II, visual artists particularly turned away from a realistic presentation of emotional and psychic experiences. "In the early 'fifties, a growing number of intellectuals consciously espoused indifference as a virtue, as the correct way to deal with an uncertain world."[6] Artists such as Marcel Duchamp, John Cage, Merce Cunningham, Robert Rauschenberg, and Jasper Johns created a cult of indifference; their ideology as a group coalescing during the terrible McCarthy period. Not only were the 'fifties uncertain, they were also terrifying and terrorizing, with the result that many writers, musicians, and dancers withheld their opinions from their work, declared themselves "alienated," and turned to formalistic inquiries or an exploration of the materials of their art. Their creative process assumed a transcendence as they chose the Zen "anything can follow anything," or chance, approach to composition. Despite Tudor's own interest in Zen, he continued to choreograph in a systematic way and never espoused an aesthetic that omitted emotional and human dimensions.

Tudor's work deliberately contradicted the fashionable stance of "indifference," and this put Tudor out in the cold. Indeed, he was stigmatized as passé, especially in New York, where among modern choreographers it was almost hieratically *de rigueur* to let the movement technique speak for itself. A little later, when dances took on a political and anti-Vietnam attitude, Tudor was still the outsider. His political ballet, *Echoing of Trumpets,* had been composed too early, and, besides, was about the "wrong" war, though he quite rightly argued that the work had larger and more universal implications.

When Mikhail Baryshnikov assumed the mantle of artistic director of American Ballet Theatre in 1980, he pledged his dedication to Antony Tudor's choreography, calling Tudor the "artistic conscience" of American Ballet Theatre. American Ballet Theatre has tried valiantly to be faithful to that conscience. Yet the same issues about the datedness of Tudor's ballets continue to surface. Some time ago, Joan Acocella criticized them for being sentimental and overly literary. She alleged that they were passé because he "fingerpoints" with his "I know what you're thinking subject matter."[7] She also cautioned the directors of A.B.T. not to commission any more romantic/narrative/psychological works and praised Baryshnikov for polishing the image of the company with a snappy corps de ballet and a challenging repertoire of Twyla Tharp, Paul Taylor, Merce Cunningham, and Mark Morris.

Tudor's conception of ballet defies such easy and ultimately false evaluations. His understanding of ballet technique transcends the 1990s' formalistic, rapid-energy, quirky attitude toward movement and hit-it, stomping-on-the-beat musicality. (Although one would never accuse Cunningham, Taylor, Tharp, or Morris of such failings.) The most recent successful revival of *Undertow,* staged for A.B.T. by Sallie Wilson in 1992, confirms the continuing power of Tudor's dramatic ballets. Paradoxically, one of the reasons for its success may be the fact that it was dropped from the repertoire in 1978. In consequence, the 1992 production was given a fresh approach and new sense of purpose that some of Tudor's ballets have not yet received.

Nowadays, the scenarios of Tudor's ballets may seem to lack drive and authenticity, not from any innate deficiency but because the dancers are no longer trained to understand character motivation or sensitized to the psychological value of an understated gesture. Although Acocella did praise the Italian ballerina Carla Fracci's "ravishingly rendered" performance of Caroline in *Jardin aux Lilas* (in the May 1990 American Ballet Theatre production at the Metropolitan Opera House), most dancers today lack teachers who will work through the context of a movement, who will indulge in the time-consuming exercises that help reveal a character's sensibility. Though we often see exciting revivals of old-fashioned stories in operas and plays, with singers or actors who infuse their roles with complex meanings, it is sadly uncommon to see a Tudor ballet that stuns us with its psychological depth. One feels that dancers today simply do not understand his performance style.[8]

It is not the dancer's fault that ballet classes avoid what cannot be judged technically. A common criticism is that everything seems geared to the classroom as an end in itself. Donald Mahler has said of Tudor: What one sees onstage is what one sees in the classroom, only somehow it has become even less interesting. Tudor's approach was the opposite.

> He brought the stage to the classroom and in so doing was able to enrich what happened onstage: He emphasized quality of movement and musicality. His art was based on an insightful understanding of human nature and he always probed and sought ways to make dancing expressive and communicative, not in superficial ways where acting was superimposed on academic movements, but rather in the deepest way, from the inside out.[9]

When assessing Tudor's place in the history of twentieth-century dance, one must not forget that he clearly influenced Jerome Robbins, Robert Joffrey, Eliot Feld, and Jiří Kylián. These choreographers have often acknowledged their indebtedness to Tudor, as mentor and master. *Dances at a Gathering* (1969) reveals Robbins's choreographic responses to Chopin mazurkas, études, etc., reminding one of Polish folk dances although rooted in classical technique, much like Tudor's evocative and stylistic use of music in *Dark Elegies* and *Echoing of Trumpets.* Similar to *Dim Lustre* and parts of *Jardin aux Lilas,* Joffrey's *Remembrances* (1973), to the music of Wagner's Wesendonck poems, explores the subtle episodic moods of memory with fluid intermixings of romantically happy and sad love coupling. Feld's *Intermezzo* (1969), to music by Brahms, with its inventive and complex classical patterning, owes its nostalgic aura to many of Tudor's ballets, where dancing becomes the means by which different languages of love are spoken. The Czechoslovakian choreographer Jiří Kylián makes ballets, especially *The Overgrown Path* (1981), that convey emotional values of real dramatic impact. The movement style as well as the smooth and liquid technique are also characteristic of Tudor. All of these works would not have been created without Tudor, as it was he who revealed that lyrical movement is not incompatible with the depiction of reality.

Since the death of Tudor, the estate, whose executrix is Sally Brayley Bliss, has given permission to many companies to perform his works, far more than during his lifetime. The people who have inherited the right to stage and direct Tudor's ballets, such as Sallie Wilson, Airi Hynninen, Maria Grandy, and Donald Mahler, for instance, often bemoan the difficulty of keeping the Tudor oeuvre vital and sensitive to changes in audience attitudes while remaining true to its integrity. One hopes that new administrations of the American Ballet Theatre as well as other ballet companies will carry on Tudor's work. From the Pacific Northwest Ballet to the Royal Winnipeg Ballet, from the Tulsa Ballet to the Ballet de Lille in France, there is now a resurgence of interest in using Tudor's works to provide regional repertoires

243

with story ballets of deep meaning and enduring choreographic values that can be used to train dancers as actors onstage. There is a challenge to future Tudor dancers who take themselves too seriously or think of themselves too highly. They should heed Tudor's words: "I think being on stage is a golden opportunity to escape from our own miserable selves. Besides, stardom comes and goes, doesn't it?"[10]

Despite the many problems encountered in remounting his ballets, they repay our close attention. When reflecting on his teaching and choreography, we realize that Tudor always led us back to the heart, linking the cool rationales of geometric patterns and shapely movements with the sentiments of passion, love, and violence. He inaugurated a bold choreographic lyricism that combined delicate artistry with attacks against conventional moral standards. He appealed to our awareness of the Freudian subconscious, to the unspoken impulses toward violence and sensuality that social standards either deny or repress. The powerful poetry of his ballets abounds in paradoxes: compassion for those who suffer and relentless irony for those who think too well of themselves.

Appendix A
Tudor's Choreography

Ballets

Titles preceded by an asterisk are fully discussed in the text.

Abbreviations:

Scen.—scenario; mus.—music; arr.—arranged; orch.—orchestration; dec.—decor; cos.—costumes; prem.—premiere; chor.—choreography

1931
Cross-Garter'd
Scen. after William Shakespeare; mus. Frescobaldi (selected organ works); dec. Pamela Bocquet; cos. after Burnacini.
Prem. November 12, Mercury Theatre, Ballet Club.
Cast: Olivia/Maude Lloyd; Maria/Prudence Hyman; Malvolio/Antony Tudor; Sir Toby/Walter Gore; Sir Andrew/Rollo Gamble; Fabian/William Chappell; Attendants/Elisabeth Schooling, Betty Cuff.
In the program Tudor is listed as stage manager and Hugh Bradford as the pianist.

1932
Mr. Roll's Quadrilles (also called *Mr. Roll's Military Quadrilles*) Mus. traditional; cos. Susan Salaman.
Prem. February [4? 11?], Mercury Theatre, Ballet Club.

Cast: The Leaders/Prudence Hyman, Antony Tudor; Pas de Trois Maude Lloyd, Elisabeth Schooling, Betty Cuff.
Presented on the Thursday Evenings at Nine series.

Constanẓa's Lament
Inspired by Leonide Massine's "The Good-Humoured Ladies," mus. Domenico Scarlatti.
Prem. February 4 or 11, Mercury Theatre, Ballet Club.
Cast: Diana Gould.

Lysistrata
Mus. Serge Prokofiev (piano pieces from Op. 2, 3, 12, and 22, and from 2nd and 4th Sonatas); dec. and cos. William Chappell, scen. after Aristophanes.
Prem. March 20, Mercury Theatre, Ballet Club.
Cast: Myrrhina, a young mother/Alicia Markova; Cinesias, her husband/ Walter Gore; Lysistrata/Diana Gould; Husband to Lysistrata/Antony Tudor; Calonice/Prudence Hyman; Husband to Calonice/William Chappell; Lampito/Andrée Howard; Other Athenian Women/Elisabeth Schooling; Handmaid to Myrrhina/Susette Morfield.

Adam and Eve
Mus. Constant Lambert; dec. and cos. John Banting.
Prem. December 4, Adelphi Theatre, Camargo Society.
Cast: Adam/Anton Dolin; Eve/Prudence Hyman; The Serpent/Antony Tudor; The Angel/Natasha Grigorova; Seraphim/Ciceley Grave, Felicity Andreae; Fowls of the Air/Peggy Van Praagh, Susette Morfield, Christine Rosslyn, Molly Brown.

1933
Pavane pour une Infante Défunte—details unavailable.

Atalanta of the East
Mus. arranged by Theodore Szanto and Seelig with additional work by Constant Lambert to music based on Javanese tunes, as well as Debussy's *Pagodes*. Dec. and cos. William Chappell; cos. executed by Mills.
Prem. May 7, Mercury Theatre, Ballet Club.
Cast: Sita/Pearl Argyle; Vikram (a wooer)/Hugh Laing; The Goddess/Diana Gould; Attendant deities/Tamara Svetlova, Yvonne Madden; King/Antony Tudor; Queen/Anna Brunton; Apsari/Elisabeth Schooling, Betty Cuff, Elizabeth Ruxton, Susette Morfield, Joan Benthall. At the piano, Charles Lynch.

1934
Paramour
Mus. William Boyce (selections from the Symphonies, etc.); arr. Constant Lambert; cos. John Lear.
Prem. February 20, Town Hall, Oxford, for the Oxford University Dramatic Society, for their production of Christopher Marlowe's *Dr. Faustus.*
Cast: Alexander/Walter Gore; Thais, his paramour/Diana Gould; Darius/ Hugh Laing.
Further information unavailable.

The Legend of Dick Whittington
Mus. Martin Shaw; dec. Eric Newton; cos. Stella Mary Pearce.
Prem. May 28, Sadler's Wells Theatre, as an interlude in *The Rock,* a pageant play by T. S. Eliot.
Cast: The Cat/Patricia Shaw Page; Dick/Joan Birdwood-Taylor; The Cook/ Raymonde Seton; A Sailor/Gladys Scott; The King of Barbary/Eileen Harris, Phyllis Bull; Alice/Betty Percheron; Bellringers.

The Planets
Mus. Gustav Holst; dec. and cos. Hugh Stevenson.
Prem. Oct. 28, Mercury Theatre, Ballet Club.
Cast: Mortals born under Venus—The Lovers/Pearl Argyle, William Chappell; The Planet Venus/Maude Lloyd with Tamara Svetlova, Nan Hopkins, Joan Lendrum, and Margaret Hawkins. Mortal born under Mars—The Fighter/Hugh Laing; The Planet Mars/Diana Gould with Elisabeth Schooling, Susette Morfield, Peggy van Praagh, and Isobel Reynolds. Mortal born under Neptune—The Mystic/Kyra Nijinsky; The Planet Neptune/Antony Tudor with Nan Hopkins and Joan Lendrum
    Five years after the premiere, Tudor mounted a fresh production of *The Planets* for his newly created London Ballet on January 23, 1939, with the addition of another scene, "Mercury," created for Guy Massey as the Planet, with Peggy von Praagh as the Mortal.

1935
The Descent of Hebe
Mus. Ernest Bloch (Concerto Grosso); dec. and cos. Nadia Benois
Prem. April 7, Mercury Theatre, Ballet Club.
Cast: Hebe/Pearl Argyle; Mercury/Hugh Laing; Night/Maude Lloyd; Hercules/Antony Tudor; Attendants on Jupiter/Susette Morfield, Ann Gee, and Horah Whitworth; Attendants on Night/Tamara Svetlova, Joan Lendrum,

and Cecily Robinson. The Children/Bridget Kelly, Bitten Nissen, Cyril Hay, and Paul Forbes.

A string quartet accompanied the first performance.

1936

*Jardin aux Lilas
Mus. Ernest Chausson (Poéme); dec. and cos. Hugh Stevenson; cos. executed by Pat McCormick.
Prem. January 26, Mercury Theatre, Ballet Rambert.
Cast: Caroline/Maude Lloyd; Her Lover/Hugh Laing; The Man She Must Marry/Antony Tudor; An Episode in His Past/Peggy van Praagh; The Sister/Elisabeth Schooling; The Sailor/Frank Staff; The Young Cousin/Ann Gee; The Soldier/Leslie Edwards; A Friend of the Family/Tamara Svetlova. Violinist; Jean Pougnet.
Interesting to note is the fact that *Jardin aux Lilas* played on alternate days at the Mercury with T. S. Eliot's *Murder in the Cathedral.*

1937

*Dark Elegies
Mus. Gustav Mahler (Kindertotenlieder); dec. and cos. Nadia Benois.
Prem. February 19, Duchess Theatre, Ballet Rambert.
Cast: I/Peggy van Praagh with chorus; II/Maude Lloyd and Antony Tudor; III/Walter Gore with Tudor, John Byron, and chorus; IV/Agnes de Mille; V/Hugh Laing with chorus. Chorus: Daphne Gow, Ann Gee, Patricia Clogstoun, Beryl Kay, Celia Franca. Singer: Harold Child, Baritone.

Suite of Airs—details unavailable

Gallant Assembly
Mus. Giuseppe Tartini (Cello Concerto in D); dec. and cos. Hugh Stevenson.
Prem. June 14, Playhouse Theatre, Oxford Dance Theatre.
Cast: Aristocrats in Love/Agnes de Mille, Peggy van Praagh, Phyllis Bidmead, Victoria Fenn, Antony Tudor, Hugh Stevenson; Hired Performers/ Margaret Braithwaite, Hugh Laing. Cellist: Muriel Taylor.

1938

Seven Intimate Dances
Curtain-raiser to Gogol's play *Marriage.*
The Hunting Scene: mus. J. C. Bach; chor. Tudor. *Joie de Vivre:* mus. Jacques Offenbach, Johann Strauss, Weston; chor. Tudor. All other dances chor. Agnes de Mille.

Prem. June 15, Westminster Theatre.

Cast: Agnes de Mille, Antony Tudor, Hugh Laing, Charlotte Bidmead, Therese Langfield.

First Series:

May Day—Beethoven, *The Hunting Scene, Stagefright*—Delibes, *Pavane*— from the film *Romeo & Juliet, Ouled Nail*—Kurdish tunes. *Joie de Vivre, The Parvenues*—Waldteufel, Strauss.

Second Series (only six dances):

Burgomasters Branle, Joie de Vivre, Audition, Hymn, The Hunting Scene, The Parvenues

Third Series:

Pavane, Stagefright after Degas, Joie de Vivre, The Parvenues, The Hunting Scene, Hymn, Judgment of Paris (*see below*).

*Judgment of Paris

Mus. Kurt Weill (selections from *Die Dreigroschenoper*); dec. and cos. Hugh Laing.

Prem. June 1, Westminster Theatre, London special company.

Cast: Juno/Therese Langfield; Venus/Gerd Larsen; Minerva/Charlotte Bidmead; with Tudor and Laing.

Soirée Musicale

Mus. Gioacchino Rossini; arr. Benjamin Britten; cos. Hugh Stevenson.

Prem. November 26, Palladium Theatre, Imperial Society of Teachers of Dancing.

Cast: Canzonetta/Gerd Larsen and Hugh Laing; Tirolese/Maude Lloyd and Antony Tudor; Bolero/Peggy van Praagh; Tarantella/Monica Boam and Guy Massey.

Re-created Toynbee Hall, December 12, 1938, with the same cast.

*Gala Performance

Mus. Sergei Prokofiev (Piano Concerto No. 3, first movement; Classical Symphony); dec. and cos. Hugh Stevenson.

Prem. December 5, Toynbee Hall, London Ballet.

Cast: La Reine de la Danse (from Moscow)/Peggy van Praagh; La Déesse de la Danse (from Milan)/Maude Lloyd; La Fille de Terpsichore (from Paris)/Gerd Larson; Cavaliers/Antony Tudor, Guy Massey, Hugh Laing; Coryphées/Monica Boam, Rosa Vernon, Sylvia Hayden, Charlotte Bidmead, Susan Reeves, Katharine Legris; Conductor/Richard Paul; Dresser/ Therese Langfield. Pianofortes played by Hans Gellhorn and Dorothy Moggridge.

1940
Goya Pastoral
Mus. Enrique Granados (Piano Pieces, orch. Harold Byrns); dec. and cos. Nicholas de Molas.
Prem. August 1, Lewisohn Stadium, New York Ballet Theatre.
Cast: Majas/Nora Kaye, Alicia Alonso; Escorts/Donald Saddler, Jerome Robbins; Marchesa/Lucia Chase; Young Man/Hugh Laing; Ladies in Waiting/Tania Dokoudovska, Miriam Golden, Maria Karniloff, Kirsten Valbor; Young Men/Fernando Alonso, John Kriza, David Nillo, Oreste Sergievsky; Fools/Leon Danielian, Dimitri Romanoff, Mimi Gomber, Olga Suarez; H. E., the Ass/Eugene Loring; Nobleman/Antony Tudor; Maiden Carrying a Basket of Grapes/Tilly Losch.
A program from the Chicago Opera House performance, November 24, 1940, still calls the ballet *Goyescas*.

1941
Time Table
Mus. Aaron Copland (Music for the Theatre).
Prem. May 29, Little Theatre of Hunter College, New York/The American Ballet.
Cast: The Station Master/Zachary Solov; The Girl/Gisella Caccialanza; Her Boy Friend/Lew Christensen; High School Girl/Lorna London; High School Boy/Charles Dickson; Three Young Girls/Beatrice Tompkins, Mary Jane Shea, June Graham; Two Marines/John Kriza, Newcomb Rice; Lady with a Newspaper/Georgia Hiden; Soldier/Jack Dunphy.
In a program from the South American tour (Paraguay), the notes indicate that "El ballet *Despedida* trata de evocar el periodo de la postguerra." On the same program were *Concerto Barocco* and *The Cave of Sleep* by George Balanchine, and *Charade* by Lew Christensen.

1942
*Pillar of Fire
Mus. Arnold Schoenberg (Verklaerte Nacht); dec. and cos. Jo Mielziner.
Prem. April 8, Metropolitan Opera House, Ballet Theatre.
Cast: Eldest Sister/Lucia Chase; Hagar/Nora Kaye; Youngest Sister/Annabelle Lyon; Friend/Antony Tudor; Young Man from the House Opposite/Hugh Laing; Lovers in Innocence/Maria Karnilova, Charles Dickson, Jean Davidson, John Kriza, Virginia Wilcox, Wallace Seibert, Jean Hunt, Barbara Fallis; Lovers in Experience/Sono Osato, Rosella Hightower, Muriel Bentley, Jerome Robbins, Donald Saddler, Frank Hobi; Maiden

Ladies Out Walking/Galina Razoumova, Roszika Sabo. Conductor: Antal Dorati.

1943

*The Tragedy of Romeo and Juliet
Scen. Antony Tudor, based on the play by Shakespeare; mus. Frederick Delius (Eventyr, Over the Hill and Far Away, Brigg Fair, The Walk to Paradise Garden, and Prelude to Irmelin, arr. Antal Dorati).
Prem. April 6 (incomplete), April 10 (complete), Metropolitan Opera House, Ballet Theatre.
Cast: Montague/Borislav Runanine; Capulet/John Taras; Romeo/Hugh Laing; Mercutio/Nicolas Orloff; Friar Laurence/Dimitri Romanoff; Paris/ Richard Reed; Lady Montague/Miriam Golden; Lady Capulet/Galina Razoumova; Juliet/Alicia Markova; Rosaline/Sono Osato; Nurse/Lucia Chase; Kinsmen of the Montagues and Capulets, Ladies at the Ball, Townspeople, Attendants/Corps de Ballet.

*Dim Lustre
Mus. Richard Strauss (Burleske for Piano and Orchestra); dec. and cos. Motley.
Prem. October 20, Metropolitan Opera House, Ballet Theatre.
Cast: Waltzing Ladies and Their Partners/Barbara Fallis and Harold Lang, June Morris and Kenneth Davis, Roszika Sabo and Fernando Alonso, Mimi Gomber and John Taras, Mary Heater and Alpheus Koon; The Lady with Him/Nora Kaye; The Gentleman with Her/Hugh Laing; A Reflection/Muriel Bentley; Another Reflection/Michael Kidd; It Was Spring/John Kriza; Who Was She?/Janet Reed, Albia Kavan, Virginia Wilcox; She Wore a Perfume/ Rosella Hightower; He Wore a White Tie/Antony Tudor.

1945

*Undertow
Scen. Antony Tudor after a suggestion by John Van Druten; mus. William Schuman; dec. and cos. Raymond Breinin.
Prem. April 10, Metropolitan Opera House, Ballet Theatre.
Cast: The Transgressor/Hugh Laing; Cybele/Diana Adams; Pollux/John Kriza; Volupia/Shirley Eckl; Aganippe/Patricia Barker; Sileni/Regis Powers, Stanley Herbertt; Satyrisci/Michael Kidd, Fernando Alonso, Kenneth Davis, Roy Tobias; Nemesis/Roszika Sabo; Polyhymnia/Lucia Chase; Pudicitia/ Cynthia Risely; Ate/Alicia Alonso; Hymen/Dick Beard; Hera/Janet Reed; Bacchantes/Marjorie Tallchief, June Morris, Mildred Ferguson; Medusa/ Nana Gollner.

1948
Shadow of the Wind
Mus. Gustav Mahler (Das Lied von der Erde); dec. and cos. Jo Mielziner.
Prem. April 14, Metropolitan Opera House, Ballet Theatre.
Cast: I. Six Idlers of the Bamboo Valley/Igor Youskevitch, Hugh Laing, Dimitri Romanoff, and ensemble; II. The Abandoned Wife/Alicia Alonso, John Kriza, Mary Burr; III. My Lord Summons Me/Ruth Ann Koesun, Crandall Diehl, and ensemble; IV. the Lotus Gatherers/Diana Adams and ensemble; Zachary Solov and ensemble; V. Conversation with Winepot and Bird/ Hugh Laing; VI. Poem of the Guitar/Nana Gollner, Hugh Laing, Dimitri Romanoff, Muriel Bentley, Barbara Fallis, Igor Youskevitch, and ensemble; Singers/Robert Bernauer and Louise Bernhardt.

1949
The Dear Departed
Mus. Maurice Ravel.
Prem. July 15, Jacob's Pillow.
Cast: Diana Adams and Hugh Laing.

1950
Nimbus
Mus. Louis Gruenberg (Violin Concerto); dec. Oliver Smith; cos. Saul Bolasni.
Prem. May 3, Center Theatre, Ballet Theatre.
Cast: Dreamer/Nora Kaye; Dream/Diana Adams; Dream-Beau/Hugh Laing; Irrelevancies/Virginia Barnes, Jacqueline Dodge, Dorothy Scott, Jenny Workman, Eric Braun, Jimmy Hicks, Fernand Nault, Holland Stoudenmire; Nightbirds/Mary Burr, Barbara Lloyd, Charlyne Baker, Virginia Barnes, Jenny Workman, Eric Braun, Jimmy Hicks, Fernand Nault, Holland Stoudenmire.
Violin Solo: Broadus Erie. Conductor: Max Goberman.
   The music was originally commissioned by Jascha Heifetz and performed by him with the Philadelphia Orchestra in 1944. The opening-night program also featured *Princess Aurora, Designs with Strings,* and *Interplay.*

1951
Lady of the Camellias
Mus. Giuseppe Verdi (selected by Tudor from *Nabucco, Sicilian Vespers, Macbeth,* and *I Lombardi*); dec. and cos. Cecil Beaton.
Prem. February 28, New York City Center, New York City Ballet.
Cast: Prudence/Vida Brown; Marguerite Gautier/Diana Adams; M. le Comte

de N./Brooks Jackson; Armand Duval/Hugh Laing; Armand's Father/ Antony Tudor.

This ballet was presented on the same program as *Card Game, Sylvia: Pas Deux,* and *Bourrée Fantasque.* This production was made possible through the courtesy of Ballet Associates in America, Inc.

Les Mains Gauches
Mus. Jacques Ibert.
Prem. July 20, Jacob's Pillow.
Cast: Zebra Nevins, Sallie Wilson, Marc Hertsens.

Ronde du Printemps
Mus. Erik Satie.
Prem. August 1, Jacob's Pillow.
Cast: Francine Bond, Harvey Jung, Marc Hertsens, Jack Monts, Zebra Nevins, Adelino A. Palomonos, Sallie Wilson.

1952
La Gloire
Mus. Ludwig van Beethoven (Overtures: *Egmont, Coriolanus,* and *Leonora* No. 3); dec. Gaston Longchamp; cos. Robert Fletcher.
Prem. February 26, New York City Center, New York City Ballet.
Cast: "La Gloire"/Nora Kaye; Handmaidens to Lucretia/Barbara Walczak, Beatrice Tompkins; Sextus Tarquinius/Francisco Moncion; The Pleiads/ Doris Breckenridge, Edith Brozak, Arlouine Case, Kaye Sargent, Tomi Wortham, Gloria Vauges; Orion/Jacques d'Amboise; Artemis/Una Kai; Hippolytus (Stepson to Phaedra)/Hugh Laing; Ophelia/Breckenridge; The Players/Walczak, Walter Georgov, Stanley Zompakos; Hamlet's Mother/ Tompkins; Hamlet's Stepfather/Moncion; Laertes/Laing; The Dancer in Gray/Diana Adams. Lightning by Jean Rosenthal.

Other ballets on the opening-night program were *Four Temperaments, Symphony in C,* and *À la Françaix.*

*Trio Con Brio
Mus. Mikhail Glinka; chor. Vispitin; cos. Marie Nepo, Adolphine Rott.
Prem. June 27, Jacob's Pillow, Lee, Massachusetts.
Cast: Introduction, Mazurka, and Gracioso: Tatiana Grantzeva, Nicholas Polajenko, Ralph McWilliams. Allegro—Variation I: Polajenko; Variation II: Grantzeva; Variation III: McWilliams. Lezginka; all three dancers.

"Vispitin" is a pseudonym for Tudor, who did not want the work to bear his name because it was not in his usual style.

1953
Exercise Piece
Mus. Ariaga y Balzola (String Quartet No. 2 in A).
The quartet is divided into three sections, *Allegro con brio,* Theme & Variations: *Andante,* and *Andante ma non troppo: Allegro.*
Prem. May 7, Concert Hall, Juilliard School of Music.
Cast: Students of Juilliard.
Preceded by a lecture by Tudor, "Let's Be Basic." Doris Humphrey presented a lecture-demonstration on the same program. The ballet was approximately twenty minutes long.

*Little Improvisations
Mus. Robert Schumann (Kinderscenen: I. From Foreign Lands and People; II. Funny Story; III. Blindman's Buff; IV. Suppliant Child; V. Perfect Happiness; VI. A Great Event; VII. Reveries; VIII. Ride a Cock Horse; IX. The Poet Speaks [King and Queen and Finale]).
Prem. August 28, Jacob's Pillow.
Cast: Yvonne Chouteau, Gilbert Reed.
   There is a film of *Little Improvisations,* made at Juilliard in 1960, that runs eleven minutes, with Elizabeth Sawyer at the piano, Sirpa Jorasma and Eric Hampton dancing.

Elizabethan Dances
Mus. Orlando Gibbons "Lord Salisbury's Pavan and Galliard"; "Coronto Suite" (Anon.); "Almain," Anthony Holborne; Thomas Morley, "La Volta," set to virginals by William Byrd; "Heigh Ho for a Husband" from John Gamble's *Common Place Book;* "Canaries" (Anon.); Thomas Tomkins, "Worsters Braule."
Prem. December 7, Concert Hall, Juilliard School of Music, for "A Festival of British Music."
Cast: Students of Juilliard Dance Department.

Britannia Triumphans
Mus. William Lawes; dec. Frederick Kiesler; cos. Leo Van Witsen.
Prem. December 11, Juilliard School of Music, for "A Festival of British Music."
Cast: The Anti-Masques; Entry, Descent, and Dances of the Grand Masquers; Galatea's Nymphs/Students of Juilliard Dance Department.
Mus. "Lord Salisbury's Pavan and Galliard"; "La Rondinella" and "La Volta"; "Coronto Suite" (Anon.); "Almain" (Anthony Holborn); Canaries (Anon.); and "Worsters Braule" (Thomas Tomkins).
Prem. Dec. 7, Concert Hall, Juilliard School of Music for "A Festival of British Music."

1954
Offenbach in the Underworld; or, Le Bar Du Can-Can
Mus. Jacques Offenbach (*Gaîté Parisienne*).
Prem. May 8, Convention Hall, Philadelphia, Philadelphia Ballet Company.
Cast: Madame la Patronne/Ruth Anne Carr; Her Little Girl/Paula Mainwaring; The Painter (from Abroad)/Michael Lland; Debutante/Sylvia Kim; His Imperial Excellency (from Abroad)/Michael Lopuszanski; The Visiting Operetta Star (from Abroad)/Viola Essen; An Officer/Maurice Phillips; The Queen of the Carriage Trade/Elaine Wilson; Les Garçons, Local Ladies, Local Café Habitués, Debutantes/Corps de Ballet.

1956
Pas de Trois
Mus. Carl Maria von Weber (Overture to *Euryanthe*).
Prem. April 26, 1956, Juilliard Dance Division, Juilliard School of Music.
Cast: Caroline Bristol, Gail Valentine, Bruce Marks. The ballet's length is approximately twelve minutes.

1958
La Leyenda de Jose—details unavailable.

1959
Hail and Farewell
Mus. Richard Strauss ("Frühling," "Beim Schlafengehn," and "September," all with lyrics by Herman Hesse; and "Im Abendrot," lyrics by J. V. Eichendorff).
Prem. March 22, Metropolitan Opera House, Metropolitan Opera Ballet.
Cast: I. Festival March (Opus 1): Suzanne Ames, Meredith Baylis, Ann Etgen, Catherine Horn, Nancy King, Carole Kroon, Fronda Sobel, Mary Stone, William Burdick, Jose Gutierrez, Harry Jones, Donald Martin, David Milnes, Ron Murray, Vincent Warren, Alek Zybine; II. Serenade (Opus 7): Margaret Black, Hlenka Devon, Louellen Sibley, Ann Etgen, Fronda Sobel, William Burdick, Ron Murray, Vincent Warren; III. Four Last Songs: Fruhling/Lupe Serrano; September/Edith Jerell; Beim Schlafengehn/Audrey Keane, Hlenka Devon, William Burdick, Ron Murray, Vincent Warren; Im Abendrot/Nora Kaye. Sung by Eleanor Steber.

1960
A Choreographer Comments
Mus. Franz Schubert (Octet in F, Op. 166, D 803, 1824); chor. Tudor; lighting Thomas DeGaetani.

Prem. April 8, Concert Hall, Juilliard School of Music.

Cast

Comment I: Arabesque—A position in which the body is supported on one leg, while the other is extended in back with the arms harmoniously disposed. 587 Arabesques: Philippine Bausch, Chieko Kikuchi, Jennifer Masley, Michal Imber, Jerry King, Virginia Klein, Myron Howard Nadel, Carol Lipman, Koert Stuyf.

Comment II: Jeté—A spring from one foot to the other. 224 Jetés: Jennifer Masley, Benjamin Heller, Mabel Robinson, William Louther.

Comment III: Pas de Bourrée—Three transfers of weight from one foot to the other. Philippine Bausch and Koert Stuyf.

Comment IV: Tour—a turn. 60 turns: Carol Egan, Chieko Kikuchi, Barbara Hale, Jennifer Masley.

Comment V: Quatrième en l'air—Leg extended in front. Philippine Bausch, Mabel Robinson, Carol Egan, Benjamin Heller, Barbara Hale, Jerry King, Michal Imber, William Louther, Virginia Klein, Myron Howard Nadel, Carol Lipman, Koert Stuyf.

Comment VI: Bourrée Couru—Small running steps. Cheiko Kikuchi.

Comment VII: Petite Batterie—Small jumping steps in which the legs beat together. 597 beats: Carol Egan, Cheiko Kikuchi, Barbara Hale, Jennifer Masley.

Comment VIII: Posé—A step onto a straight leg. 65 posés: Jennifer Masley.

Comment IX: Tour—A turn. 184 turns: Carol Egan, Cheiko Kikuchi, Barbara Hale, Jennifer Masley.

Comment X: Pas de chat—Literally, step of a cat. One pas de chat: Michal Imber and Benjamin Heller.

An original Juilliard film runs twenty-three minutes.

1961
Dance Studies (Less Orthodox)
Mus. Elliot Carter (Étude III from Eight Études and Fantasy for Woodwind Quartet).
Prem. May 8, Juilliard School of Music.
Cast: The Juilliard Dance Ensemble.
A 14$\frac{1}{2}$-minute film was made on March 8, 1962, on the Juilliard stage.

1962
Passamezzi
Mus. Antonio Gardano (Passamezzi from collection of keyboard pieces entitled "intravolatura Nova de Varie Sorte di Balli").

Prem. March 8, Juilliard School of Music.
Cast: Juilliard Dance Ensemble.
A film from Juilliard runs $5\frac{1}{2}$ minutes.

Gradus ad Parnassum
Prem. March 8, Concert Hall, Juilliard School of Music.
Tudor choreographed several new dance studies in addition to re-creating *Little Improvisations* and *Trio con Brio*. Also: *Ballet I*. Mus. Antonio Gardano ("Passamezzi" from "Collection of Keyboard Pieces"). *Ballet I*. Mus. Henry Purcell (from "Musik's Hand Maid"). *Dance Studies (Less Orthodox)* Mus. Elliot Carter (from Eight Etudes and a Fantasy for Woodwind Quartet").

1963
*Fandango
Mus. Antonio Soler.
Prem. March 26, Town Hall, New York, Metropolitan Opera Ballet Studio.
Cast: Suzanne Ames, Ingrid Blecker, Nancy King, Carol Kroon, Ayako Ogawa.
There are several videos, two from Juilliard, of this fourteen-minute ballet.

*Echoing of Trumpets (orig. *Echoes of Trumpets*)
Mus. Bohuslav Martinu (Fantaisies Symphoniques); dec. Birger Bergling.
Prem. Sept. 28, Royal Opera House, Stockholm, Sweden, Royal Swedish Ballet.
Cast: Gerd Andersson, Catharina Ericson, Viveka Ljung, Kerstin Lust, Hervor Sjostrand, Kari Sylwan, Annette Wiedersheim-Paul, Mario Mengarelli, Jacques De Lisle, Eki Eriksson, Ulf Gadd, Nils Johansson, Nisse Winquist, Richard Wold, Svante Lindberg.

1966
Concerning Oracles
Ballet in three scenes (scene 2, *Les Mains Gauches,* 1951, *q.v.*).
Mus. Jacques Ibert (Suite Elisabéthaine, Capriccio, Divertissement; originally composed for *An Italian Straw Hat,* based on the farce by Eugène Labiche); dec. and cos. Peter Harvey.
Prem. March 27, Metropolitan Opera House.
Cast: Teller of Fortunes/Nira Paaz; Gypsies/Donald Mahler, Jan Mickens.
Scene I. *Regard sur un Crystal* (The Orb, The Chaplet, and The Skull): Edith Jerell and Susana Aschieri, Nicolyn Emanuel, Sylvia Grinvald, Rhodie Jorgenson, Janet Morse, Sharon O'Connell; Robert Davis and Martin Fried-

man, William Maloney, David Milnes, Franklin Yezer, Susana Aschieri, Sylvia Grinvald, Janet Morse, Sharon O'Connell, Josef Gregory, Howard Sayette and Sally Brayley.

Scene II. *Les Mains Gauches* (The Rose and the Noose): Carolyn Martin and Ivan Allen.

Scene III. *L'Arcane* (*Cards of the Tarot*—The Fool, Love, Marriage, Strife, The Woman of Mystery): Lance Westergard (debut), Susana Aschieri, Nicolyn Emanuel, Janet Morse; Sylvia Grinvald, Howard Sayette, Robert Davis, Josef Gregory, William Maloney, Franklin Yezer, and Sally Brayley. Singers: Mary Ellen Pracht, Carolotta Ordassy, Marcia Baldwin, Nedda Casei.

1967

*Shadowplay

Scen. based on Rudyard Kipling's *The Jungle Book;* mus. Charles Koechlin (Les Bandar-Log and La Course du Printemps); dec. and cos. Michael Annals.

Prem. January 25, Royal Opera House, London, Royal Ballet.

Cast: The Boy with Matted Hair/Anthony Dowell; The Penumbra: Arboreals/Kenneth Mason, Lambert Cox, Keith Martin, Geoffrey Cauley, Peter O'Brien, Donald Kirkpatrick, Ann Howard, Marilyn Trounson, Frank Freeman; Aerials/Ann Jenner, Jennifer Penney, Deirdre O'Conaire, Rosalind Eyre, Christine Beckley, Diana Vere; Terrestrial/Derek Rencher; Celestials/Merle Park with David Drew, Paul Brown.

1968

Knight Errant

Mus. Richard Strauss (*Le Bourgeois Gentilhomme* and Prelude to *Ariadne auf Naxos*); dec. and cos. Stefanos Lazaridis.

Prem. Nov. 25, Opera House, Manchester, England, Royal Ballet of London. Two different casts: Chevalier d'Amour (A Gentleman of Paris)/Hedrik Davel, David Wall; A Woman of Consequence (Jane Landon) Caroline Southam; Gentlemen of Standing/Spencer Parker, Peter Fairweather, Michael Ingleton, Victor Kravchenko; Ladies of Quality/Jacqueline Lansley, Kathleen Denley, Vicki Karras, Susan Lawe, Brigid Skemp; Ladies of Position—Alfreda Thorogood/Sally Inkin, Margaret Barbieri/Sandra Conley, Elizabeth Anderton/Patricia Ruanne; Gentlemen of Means/Adrian Grater, David Gordon, Kerrison Cooke; Postillions/Sven Bradshaw, Christopher Carr, Terence Hyde, Alan Hooper, Cavan Orwell, Brian Bertscher.

The following musical sequence was listed in the program: (1) Overture to *Le Bourgeois Gentilhomme* (*BG*), (2) The Dinner (*BG*), (3) Tailor's

Dance (*BG*), (4) Fencing Master (*BG*), (5) Intermezzo (*BG*), (6) Finale to Act 2 (*BG*), (7) The Dinner (*BG*), (8) Prelude to *Ariadne auf Naxos,* (9) Minuet in A Major (*BG*), (10) The Dinner (*BG*), (11) Prelude to Act 2 (*BG*), and Minuet of Lully.

1969

The Divine Horsemen
Scen. based on the book by Maya Deren; mus. Werner Egk (Variations on a Caribbean Theme); dec. and cos. Hugh Laing.
Prem. August 8, Her Majesty's Theatre, Sydney, Australian Ballet.
Cast: Erzulie/Gail Stak; Ghede/Karl Welander; Damballah/Rex McNeill; and Alida Chase, Julie da Costa, Jo-Anne Endicott, Ann Fraser, Heather Macrae, Suzanne Neumann, Carolyn Rappel, Leigh Rowles, Lucyna Sevitsky, Gail Stock, Janet Vernon, Ronald Bekker, Frances Croese, Gary Heil, Graeme Hudson, Rex McNeill, Graeme Murphy, Paul Saliba, Karl Welander, Colin Peaseley.

1971

Three New Pieces for Small Groups
Prem. May 27, Juilliard Theatre, Juilliard dance students.
Presented at a private viewing, to conform with the terms of a National Endowment for the Arts grant to Tudor. The pieces were:

Sunflowers
Mus. Leos Janacek (String Quartet No. 1 [Kreutzer Sonata]).
Cast: Madeline Rhew, Airi Hynninen, Pamela Knisel, Deborah Weaver, Anthony Salatino, Larry Grenier.

Cereus
Mus. Geoffrey Gray (Quartet for Percussion, L'Inconsequenza, 1968, rev. 1970).
Cast: Jerome Weiss, Sylvia Yamada, Larry Grenier, Lance Westergard, Marc Stevens, Bonnie Oda, Angeline Wolf.
    Tudor insisted that the names of the dancers be organized in the form of an "X" on the program. There are two videotapes, both made at Juilliard, from 1971 and April 3, 1981.

Continuo
Mus. Johann Pachelbel (Canon in D).
Cast: Sirpa Jorasma, Anthony Salatino, Deborah Weaver, Raymond Clay, Madeline Rhew, Blake Brown.

1975
The Leaves Are Fading
Mus. Antonin Dvorak (from String Quartet ["Cypresses"] and other chamber music for strings); dec. Ming Cho Lee; cos. Patricia Zipprodt; lighting, Jennifer Tipton.
Prem. July 17, New York State Theatre, American Ballet Theatre.
Cast (in order of appearance): Kim Highton, Marianna Tcherkassky, Amy Blaisdell, Nanette Glushak, Linda Kuchera, Kristine Elliot, Hilda Morales, Elizabeth Ashton, Christine O'Neal, Michael Owen, Raymond Serrano, Charles Ward, Richard Schafer, Clark Tippet, Gelsey Kirkland, and Jonas Kage.

1978
The Tiller in the Fields
Mus. Antonin Dvorak (excerpts from Symphonies No. 2 and No. 6, and Ode to Nature [concert overture "In Nature's Realm," Op. 91]); dec. Ming Cho Lee; cos. Dunya Ramicova; lighting, Tom Skelton.
Prem. December 13, Kennedy Center, Washington, D.C., American Ballet Theatre.
Cast: Patrick Bissell, Nancy Collier, Cynthia Gast, Camille Izard, Lucette Katerndahl, Christie Keramidas, Lisa Lockwood, Lisa Rinehart, Kristine Soleri, Brian Adams, John Gardner, Robert La Fosse, Danilo Radojevic, Gelsey Kirkland.

Opera Ballets

1933
September 28: *Faust* (Gounod), Sadler's Wells Theatre, danced a duet with Freda Bamford; *pas de trois*/Ailene Phillips, Nadina Newhouse, Elizabeth Miller; with corps.

1934
November: *Castor and Pollux* (J. P. Rameau), Oxford University Opera Club.

1935
May 1: *La Cenerentola* (Rossini), Covent Garden. With Alicia Markova.

June 3: *Schwanda the Bagpiper* (J. Weinberger), Covent Garden.

June 4: *Carmen* (Bizet), Covent Garden.

September 23: *Koanga* (Delius), Covent Garden.

1937
January 6: *Die Fledermaus* (J. Strauss), Covent Garden Ballet under the direction of Madeline Dinely and Anthony (*sic*) Tudor; Première Danseuse, Prudence Hyman.

1939
May 1: *The Bartered Bride* (Smetana), Covent Garden. Danced by Tudor's company.

May 22: *La Traviata* (Verdi), Covent Garden. Danced by Tudor's company.

May 24: *Aïda* (Verdi), Covent Garden. Danced by Margot Fonteyn and the London Ballet.

1950
November 11: *La Traviata* (Verdi), Metropolitan Opera. Danced by Tilda Morse, Nana Gollner, and corps de ballet.

December 12: *Faust* (Gounod), Metropolitan Opera. Danced by Nana Gollner and corps de ballet.

December 20: *Die Fledermaus* (J. Strauss), Metropolitan Opera. Danced by Nana Gollner and corps de ballet.

1960
December 6: *Alcestis* (C. W. Gluck), Metropolitan Opera. Danced by the Metropolitan Opera Ballet.

December 17: *Tannhauser* (Wagner), Metropolitan Opera. Danced by the Metropolitan Opera Ballet.

Shows and Revues

1934
February 20: *Paramour*, mus. William Boyce, Town Hall, Oxford, for the Oxford University Dramatic Society production of *Dr. Faustus*.

1936
April 8: *The Happy Hypocrite,* musical play, His Majesty's Theatre, Haymarket.

Gods: Amor/Marius Goring; Mercury/Carl Harbord. Men: Lord George Hell, a wicked nobleman/Ivor Novello; Sir Follard Follard, a broken baronet/ Philip Desborough; Beau Brummel/Charles Lefeaux; The Bishop of St. Aldred's/Fewlass Llewellyn; Jenny Mere, a village maiden/Vivien Leigh. Directed by Maurice Colbourne, ballet and ensembles by Antony Tudor, orchestra under the direction of Leonard Isaacs, scenery and costumes by Motley, masks by Angus McBean.

Peggy van Praagh mentions in her book *How I Became a Ballet Dancer* that she worked with Tudor on *The Happy Hypocrite:* "I was certainly very pleased when he chose me to dance with Hugh Laing in the musical play, *The Happy Hypocrite,* for which he was arranging the dances. This was an Ivor Novello production in which Novello and Vivien Leigh starred. Hugh and I danced a Pastoral Pas de Deux and a small mimed ballet with Vivien Leigh on a stage within the stage. This was raised up and was very small, only nine feet wide and even less in depth. So we had to dance our steps very carefully or we should have fallen on to the real stage. It was excellent training in precision and control. Novello and several of the actors watched our performance from the stage every night. If we made a mistake, we couldn't hope for its not being seen! Unfortunately the play did not run for very long—a few weeks on tour and then only 6 or 7 weeks at His Majesty's Theatre in London. I was sorry when it all finished because I had enjoyed working in an atmosphere so different from a ballet company." (pp. 46, 47)

September 11: *Careless Rapture,* musical play, Theatre Royal, Drury Lane. Devised, written, and composed by Ivor Novello, dances and ensembles arranged by Joan Davis, produced by Leontine Sagan.

Tudor arranged one ballet called "The Temple of Nichaow." A list of the musical numbers indicates that the full ensemble appears in this scene: "In their dreams they live again the legend of the Temple of the Miracle where the scene takes place, and Dorothy Dickson becomes the votaress who forsook her vows for love of a Chinese Prince who was stabbed by the priests (Mr. Novello as the Prince)."

November 26. *To and Fro,* revue, Comedy Theatre.
Dances arranged by Antony Tudor, with Maude Lloyd as ballerina in "Waiting for Twilight to Fall." Lyrics by Edgar Blatt, and music by Nat Ayer, Jr. Tudor also created the dances for "Let's Take a Chance." Lloyd and Hugh Laing appear in other scenes: "Prélude," by Lord Berners; "I'm on My

Own," with Yvette Darnac; and "Symphonie Russe," to music by Prokofiev, suggested by Sophie Fedorovitch. Osbert Sitwell is the author of a
scene, "Art Knows No Nationality." There seem to be other scenes as well
that allude to the idea of war, one world, the party spirit, and so forth. Some
skits feature well-known actors such as Hermione Baddeley, Cyril Ritchard,
and Viola Tree. The ballets were designed by Sophie Fedorovitch.

1937
September 1: *Crest of the Wave,* musical play, Theatre Royal, Drury Lane.
Devised, written and composed by Ivor Novello, Lyrics by Christopher
Hassal. A cast of 150. Dances invented and arranged by Ralph Reader,
ballets arranged by Lydia Sokolova and Antony Tudor. Produced by Leon-
tine Sagan. The corps de ballet is cited as appearing in a scene called
"Versailles in Tinsel" in the dance "Tourbillon."

1939
February 22 (14?): *Johnson over Jordan,* play, New Theatre, London.
By J. B. Priestly. Principal dancer: Sepha Treble.
Tudor made the dances and pantomime.

April 3: *Nightlights,* a Suppertime Show, revue, Trocadero, London.
Produced by C. B. Cochran. Tudor arranged two dances, "The Argyll
Rooms," with costumes by Karinska, and "The Fountain" (which later
became "A Fragonard Picture"), also with costumes by Karinska. Both of
these dance scenes are based on period styles of costume and design and
movement. The archivist for the Rambert Dance Company, Jane Pritchard,
explains that "it was usual practise for the Trocadero, late-night cabaret
shows, to run for a year from April to March." (Letter to the author, April 12,
1991). It is interesting to note that ten years earlier, in 1929, Charles B.
Cochran had hired George Balanchine, Tilly Losch, and Max Rivers to do
the dances and ensembles for "Wake Up and Dream," to music and lyrics by
Cole Porter. Balanchine worked with Cochran again in 1930.

1940
February 9: *Lights Up!* revue, Savoy Theatre.
Produced by C. B. Cochran. Tudor arranged the dances in "An Old Dance
Hall" and "A Fragonard Picture."

1945
May 31: *Hollywood Pinafore,* musical, Alvin Theatre, New York.
Mus. by Arthur Sullivan; lyrics by George S. Kaufman.

The show satirized Hollywood's cash consciousness. Tudor's dances received some rave, if brief, remarks. His contribution digressed from the *H.M.S. Pinafore* score, using instead *The Mikado* and the *Yeoman of the Guard.*

November 22: *The Day Before Spring,* musical, National Theatre, New York. Book by Alan J. Lerner, mus. Frederick Loewe. The dancers were Mary Ellen Moylan and Hugh Laing.

Most reviews mentioned that Tudor's dances did not enhance the show's thin plot. In the *New York Post* (November 23), Wilella Waldorf speaks of the two dances by Tudor, in spite of Laing's performance, as not "top Tudor." Robert Coleman in the *Daily Mirror* (November 23) also criticizes Tudor for not "giving Mary Ellen Moylan and Hugh Laing a chance to display their talents. The dances are not of Antony Tudor's top shelf."

Television and Film

A handwritten letter dated December 7, 1990, from Dallas Bower, one of Tudor's major producers for the BBC-TV during the 1936 to 1939 period, explained what kind of shows were arranged for television: "I revived on TV a light entertainment forum known as intimate revue. This was popular in the London Theatre throughout the 1920s and early '30s, its chief producers being André Chavlot, Albert de Canvelle [?] and William Walker. Revues always contained ballet in one form or another. . . . I always found Tudor easy to work with, although a forceful disciplinarian inasmuch as he demanded intense concentration upon the work in hand." All television titles in the list were broadcast over BBC-TV.

1937
January 7: *Paleface,* television.
A floor show with Hermione Baddeley, Cyril Richard, Antony Tudor, Bobby Tranter's Girls. On the same program Antony Tudor and Maude Lloyd appeared in "Romeo and Juliet," music by Constant Lambert; "Sonata No. 2," music by William Boyce; "Symphony No. 1," music by William Boyce, arr. by Lambert; "Symphony No. 6," music by Boyce, arr. by Lambert; "Concerto No. 3," music by Prokofieff, accompanied by Harold Stuteley.

For information on Tudor's television ballets, see Janet Rowson Davis, "Ballet on British Television, 1933–1939," *Dance Chronicle,* vol. 5, no. 3,

1982–83; and Janet Rowson Davis, "Ballet on British Television, 1948–1949, Part 1: Company Debuts, Teleballets, Recitals," *Dance Chronicle,* vol. 15, no. 1, 1992.

February 2: *Paramour,* television.
Tudor choreographed, with producer Dallas Bower, dancers Maude Lloyd and Harold Turner. Mus. William Boyce, "Sonata No. 2," "Symphony No. 1," "Symphony No. 6," and "Romeo and Juliet." *Paramour,* based on the Thaïs story, was a reworking of Tudor's own ballet, which was first included in *Dr. Faustus,* produced by the Oxford University Dramatic Society on February 20, 1934, when he used three dancers. It was taken into the Ballet Rambert repertoire on April 22.

February 2: *Hooey,* television.
A floor show with Frances Day, Cyril Ritchard, Antony Tudor, Maude Lloyd, Bobby Tranter and His Girls.

February 2: *Siesta* (William Walton), television. Dec. and cos. by Peter Bax. Maude Lloyd and Tudor, dancers.

February 4: *Constanza's Lament,* television.
Under the aegis of Marie Rambert, with producer Stephen Thomas. A solo danced by Pamela Foster with choreography by Antony Tudor. Also on the program, Frederick Ashton's *The Passionate Pavane,* mus. D. Scarlatti.

March 2: *Fugue for Four Cameras,* mus. Bach, Fugue in D Minor, from "The Art of the Fugue."
Produced by Stephen Thomas, with Maude Lloyd dancing.
    Tudor remarked (in a taped interview about experimental film with Maya Deren in 1956), "I had to devise a dance that would fit the four-part fugue musically. But instead of another dancer joining the movement with the second voice, another camera joined the movement and then you saw the dancer (Maude Lloyd) in the center of the film open up and become a second dancer and then a third voice would come in and pick up where she started. We spent a lot of time on this, or the camera man spent a lot of time on this. It was rather exciting."

March 2: *After Supper,* television.
Revue, with Maude Lloyd, Antony Tudor, and others.

March 2: Fête d'Hébé, television.

Mus. Jean Philippe Rameau, Minuet No. 2; with Maude Lloyd, Antony Tudor, Hermione Moir, and Charles Stewart.

April 2: *Wienerblut,* mus. J. Strauss, television.
Danced by Maude Lloyd, Antony Tudor, Hermione Moir, Charles Stewart. This work is not listed in the "Programmes-As-Broadcast" list.

April 2: *The Story of the Vienna Waltz,* mus. J. Strauss, television.
Danced by Maude Lloyd, Antony Tudor, Elisabeth Schooling, John Andrewes.

April 13: *Dorset Garden,* television.
A miniature restoration revue designed to recapture the atmosphere of a well-known London playhouse of the seventeenth century, Dorset Garden Theatre.

April 13: *Suite of Airs,* mus. Purcell, selections from *The Fairy Queen,* television.
   Davis in *Dance Chronicle,* vol. 5, no. 3, p. 281, noted that Dallas Bower thought that this work was never produced on television, though it did have rehearsals for TV.
Stage premiere May 16, 1937, Mercury Theatre, Ballet Club.

June 29: *Boulter's Lock, 1908–1914,* television.
   A revue with a setting that became a favorite haunt of boating enthusiasts on the upper reaches of the Thames. Tudor and Maude Lloyd danced "The Dancing Lesson," "The Band Box Girl," and "When I Marry Amelia." In addition, Tudor danced "Captain Reginald D'Arcy of the Guards." A promotional article in *Buck's Free Press* (June 23) mentioned that the show with "water rising will be featured by the BBC. Maude Lloyd and Antony Tudor will dance." It tells the story of 1908 onward when Boulter's Lock was a microcosm of London life and fashion.

July 5: *Cabaret.* A revue, with Valerie Hobson, Richard Dolman, Ernst and Lotte Berk, Eric Wild and his Tea Timers, and Charles Zwar.

July 8: *Relâche,* mus. Satie, ballet by Francis Picabia.
Danced by Maude Lloyd, Elisabeth Schooling, Hugh Laing, Charlotte Landor, and Tudor. The show was produced by Dallas Bower. Greer Garson appeared on the same show in "How He Lied to Her Husband," by George Bernard Shaw.

July 14. 1) "En Diligence," a revue, mus. Poulenc, Waltz No. 10, danced by Tudor; 2) "Romeo and Juliet," mus. Constant Lambert, incidental music to *Romeo and Juliet,* dancers, Maude Lloyd and Antony Tudor; 3) "The Boy David," mus. William Walton, excerpts from *The Boy David,* based on James Barrie's play, danced by Maude Lloyd.

September 6: *Portsmouth Point.*
Tudor choreographed for BBC, music by William Walton, Dallas Bower was producer and dancers, Peggy van Praagh, Naomi Holmes, Bridgette Kelly, Charlotte Landor, Frank Staff, Elisabeth Schooling, Mark Baring, John Thorpe and Harry Webster.

September 14: *High Yellow,* mus. Spike Hughes, television.
Produced by Dallas Bower, with Peggy van Praagh, Charlotte Landor, Elisabeth Schooling, and Bridgette Kelly.
On June 6, 1932, Frederick Ashton had worked with Buddy Bradley on a revue using this score for the Camargo Ballet Society at the Savoy Theatre.

October 25: *Full Moon,* mus. Herbert Murrill, television.
Revue by Archie Harradine. Antony Tudor and Margaret Braithwaite dance "The Waltz" and "Full Moon Blues,"

December 31: Tudor dances with Margot Fonteyn to Erik Satie's "La Femme Rejoint son Fauteuil." On the same program a film of the De Basil Ballet was shown.

1938
January 24: *Tristan and Isolde,* Act 2. Mus. R. Wagner, television.
A masque. Mime arranged by Antony Tudor, decor and costumes by Peter Bax.
Dallas Bower recalled in his letter to me of December 7, 1990, that "My most ambitious undertaking with Tudor was the whole of Act 2 of Wagner's *Tristan and Isolde.* The principals were Olive (Oriel) Ross and Basil Bartlett, who had had no ballet or mime training specifically. The act was for just over an hour and Tudor rehearsed it for a month with super fine results." The following people performed the mime: Tristan/Basil Bartlett; Isolde/ Oriel Ross; Brangäne/Mary Alexander; King Marke/Paul Jones; Melot/Hugh Laing; Kurwenal/Anthony Hyndman.

April 5: *Wien,* television.
"A Viennese entertainment."

The ballet presented Antony Tudor as the Roué, Hugh Laing as the Hussar, Robert Dorning as the Masher, Charlotte Landor as the Dowager, Peggy van Praagh as the Maid, and Prudence Hyman as the Fashionable Girl. On the same program was the ballet "A Cavalcade of Strauss Waltzes," and the ballet "Roses of the South," by Strauss.

April 24: "Acis and Galatea," a pastoral, mus. Handel.
Ballet included Maude Lloyd, Hugh Laing, Peggy Van Praagh, Sally Gilmore, Antony Tudor, Elisabeth Schooling, Celia Franks, Charlotte Landor, and Therese Langfield.

May 11: *The Emperor Jones,* play by Eugene O'Neill, televised broadcast. Tudor played the Auctioneer, Little Formless Fears were danced by Miss Landor, Miss Langfield, and Miss Ellison, Hugh Laing was the Witch Doctor, and Guy Massey was the Prison Guard. Dallas Bower produced the program, but no director or choreographer was listed. The music was taken from various records such as "Swingtime in the Rockies," "Deep River," and "Mwana we Pfumo Chera," sung by the Chipika Singers.

May 29. *Master Peter's Puppet Show,* mus. DeFalla, television. Based on Thomas Shelton's translation of *Don Quixote* (1620), by J. B. Trend. Tudor arranged the mime and played the Scholar; Hugh Laing played the Page. The Hogarth Puppets and a monkey also appeared.

December 13. *Cinderella,* mus. Spike Hughes, opera in one act, television.
    With dancers Therese Langfield, Charlotte Bidmead, Jacqueline Saint, Susan Reeves, Antony Tudor, Hugh Laing, Guy Massey and Anthony Kelly.

1939
February 3. *Soirée Musicale,* television.
The London Ballet, directed by Antony Tudor, with Gerd Larsen, Maude Lloyd, Peggy van Praagh, Charlotte Bidmead, Rosa Vernon, Monica Boam, Hugh Laing, Guy Massey and Antony Tudor.
"Gala Performance," mus. Prokofiev's Piano Concerto, played by Hans Gellhorn and Dorothy Moggridge—two pianos. Danced by Peggy van Praagh, Maude Lloyd, Gerd Larsen, Monica Boam, Rosa Vernon, Sylvia Hayden, Charlotte Bidmead, Susan Reeves, Elizabeth Hamilton, Therese Langfield, Hugh Laing, Guy Massey, Richard Paul, and Antony Tudor. Also, mus. Prokofiev's Classical Symphony, danced by Peggy van Praagh, Maude Lloyd, Gerd Larsen, Monica Boam, Rosa Vernon, Sylvia Hayden,

Charlotte Bidmead, Susan Reeves, Elizabeth Hamilton, Hugh Laing, Guy Massey, and Antony Tudor.

February 5: *The Tempest,* William Shakespeare, mus. by Sibelius, television.
With members of the London Ballet, directed by Antony Tudor. Peggy van Praagh, Therese Langfield, Charlotte Bidmead, Elizabeth Hamilton, Hugh Laing, Guy Massey, John Regan.

April 7: *The Pilgrim's Progress* (sixteenth- and seventeenth-century music, (arr. Lionel Salter) by John Bunyan, adapted for television by H. D. C. Pepler.
Dances arranged by Antony Tudor. Therese Langfield, Elisabeth Schooling, Charlotte Bidmead appear in a large acting cast.

1953
March 2: *Ballet for Beginners: Jardin aux Lilas* with substantial extracts from *Soirée Musicale* and Part 2 of *Dark Elegies,* television.
Program initiated by Felicity Gray to promote understanding and appreciation of ballet. The series went into five editions.

1959
February 9: "The Very Eye of Night," film, Living Theatre, New York.
The program reads: "Conceived, directed, filmed and edited by Maya Deren; music by Teiji Ito. The film runs 15 minutes. Choreographic Collaboration—Antony Tudor; Assistant Director/Harrison Starr; Lighting Director/Ernst Neukanen; Dance Notator/Philip Salem; Recorded by Louis and Bebe Barron; General Assistance/M. Arsham, R. Borenstejn, H. Esterly, M. Khazoom, M. Kraft, B. Stauffacher. Made with the cooperation of the Metropolitan Opera Ballet School. Cast: Noctambulo/Philip Salem; Gemini/Richard Sandifer and Don Freisinger. Satellites: Ariel/Patricia Ferrier; Titania/Barbara Levin; Oberon/Bud Bready; Umbriel/Genaro; Uranus/Richard Englund; Urania/Rosemary Williams."

1960
Modern Ballet, film, part of "A Time to Dance" series.
Discussion with Antony Tudor about psychological insights in dance and about changes in ballet suggested by his work and by that of Fokine. Hugh Laing and Nora Kaye demonstrate with moments from Tudor's *Lilac Garden, Gala Performance, Pillar of Fire, Romeo and Juliet, Dim Lustre,* and *Undertow.*
Consult *Films on Ballet and Modern Dance,* John Mueller (New York: American Dance Guild, September 1974).

1967
Echoes (Echoing) of Trumpets, film, produced by Lars Egler in Sweden.

1971
Pillar of Fire, film, produced by Kjell Forsting in Sweden.

1974
Dark Elegies, film, produced by Lars Egler in Sweden.

1986
Antony Tudor, film documentary by Viola Eberle and Gerd Andersson. Produced in Sweden, released in the United States by Dance Horizons 1992. Early film clips and excerpts from Tudor's ballets.

1987
Eye on Dance, "Antony Tudor," television.
Produced by Celia Ipiotis and Jeff Bush.

1990
"An Evening with Tudor," television.
Dance in America, American Ballet Theatre.
Performed *Jardin aux Lilas* and *Dark Elegies.*

Appendix B
Notated Dances

Cereus (Grey)
      4 women, 5 men; 12 minutes
      Notator: Muriel Topaz, 1972

A Choreographer Comments: Jetés, Posés,
      Pas de Bourrée, Turns (Schubert)
      5 women, 3 men
      Notators: Sheila Zatroch and Margaret Cicierska,
      1963–64

Continuo (Pachelbel)
      3 women, 3 men; 7 minutes
      Notator: Muriel Topaz, 1976

Dance Studies (Less Orthodox): excerpts (Carter)
      1 woman, 2 men
      Notator: Kelly Hogan, 1963

Dark Elegies (Mahler)
      8 women, 4 men; 25 minutes
      Notator: Airi Hyninnen, 1980

Dim Lustre (Strauss)
      6 and 5 women, 4 and 5 men; ca. 20 minutes
      Notator: Airi Hyninnen, 1984–86

Echoing of Trumpets (Martinu)
    Notator: Virginia Doris, 1990 (in progress)

Elizabethan Suite: Excerpts (Gibbons)
    1 or more women, 1 or more men
    Notators: Patricia Sparrow, Lucille Badda, and
    Charles Wadsworth (supervised by
    Ann Hutchinson), 1954

Fandango (Soler)
    5 women; 12 minutes
    Notators: Christine Clarke Smith and Muriel Topaz,
    1970; rev. Muriel Topaz, 1982

Gala Performance (Prokofiev)
    3 women, 1 and 6 men; 10 corps; 25 minutes
    Notator: Leslie Rotman, 1988

Jardin aux Lilas (Chausson)
    6 women, 6 men; 18 minutes
    Notator: Muriel Topaz, 1967; Airi Hynninen, 1981

Judgment of Paris (Weill)
    3 women, 2 men; ca. 13 minutes
    Notator: Airi Hynninen, 1982–89

The Leaves Are Fading (Dvorak)
    9 women, 6 men, 1 walkon; 32 minutes
    Notator: Airi Hynninen, 1975

Little Improvisations (Schumann)
    1 woman, 1 man; 12 minutes
    Notators: Carol Wolz, 1961; rev. Muriel
    Topaz and Lynne Weber, 1972–74;
    rev. Els Grelinger, 1990

Offenbach in the Underworld (Offenbach)
    15 women, 7 men, 40 minutes
    Notator: Lynne Weber, 1975; Leslie Rotman, 1984–85

Pas de Trois (Weber)
> 2 women, 1 man, ca. 9 minutes
> Notator: Caroline Bristol, 1956

Pillar of Fire (Schoenberg)
> 13 women, 8 men; 31 minutes
> Notator: Airi Hynninen, 1982

Romeo and Juliet (Delius) (notation incomplete)
> large cast
> Notator: Airi Hynninen, 1975–76

Shadowplay (Koechlin)
> 9 women, 11 men; 27 minutes
> Notator: Judith Siddall, 1984–85

Soirée Musicale (Rossini)
> 6 women, 3 men; 12 minutes
> Notators: Ann Hutchinson, 1962; rev.
> Airi Hynninen, Muriel Topaz,
> and Rochelle Zide, 1975

Sunflowers (Janacek)
> 4 women, 2 men; 23 minutes
> Notator: Muriel Topaz, 1972

The Tiller in the Fields (Dvorak)
> 9 women, 7 men; 26 minutes
> Notator: Airi Hynninen, 1978

Undertow (Schuman)
> 12 women, 9 men; 45 minutes
> Notators: Ray Cook, 1973 (incomplete);
> Airi Hynninen, 1978–83

Appendix C
Tudor's Roles

His own choreography, unless otherwise noted.

1929
December 31. A Slain Lover, *Cupid and Death* (chor. Penelope Spencer). *Dido and Aeneas*—details unavailable.

1931
November 12. Malvolio, *Cross-Garter'd.*
Devember 31. Dancer, *The Tartans* (chor. Frederick Ashton).

1932
February. A Leader, *Mr. Roll's Quadrilles.*
February. The Trainer, "Le Boxing" from *Sporting Sketches* (chor. Susan Salaman); Edwin Morris, *The Lord of Burleigh* (chor. Frederick Ashton).
March. *The Dancing Times* mentions Tudor doing a Butterfly Ballet with the actress/singer Gina Malo.
March 11. Warrior, *The Enchanted Grove* (chor. Rupert Doone).
March 20. Lysistrata's Husband, *Lysistrata.*
October 17. Edward Gray, *The Lord of Burleigh* (chor. Frederick Ashton).
October 11. Passport Officer, *Douanes* (chor. Ninette de Valois).
November 1. Apollo, *The Origin of Design* (chor. Ninette de Valois).
December 4. Serpent, *Adam and Eve.*

1933
February 7. Rural Dancer, *The Birthday of Oberon* (chor. Ninette de Valois).

May 7. King, *Atalanta of the East.*

December 3. Mme. Récamier's Suitor, *An 1805 impression* (chor. Frederick Ashton).

September 28. *Faust* (Gounod), Sadler's Wells Theatre; danced a duet with Freda Bamford.

October 30. Man, *La Création du Monde* (chor. Ninette de Valois).

1934

April 24. The President, *Casse Noisette.*

May 28. Dancer, Interlude, *The Rock,* a pageant play, T. S. Eliot.

October 28. A Lover, "Alcina Suite," *Vauxhall Gardens* (chor. Andrée Howard)

October 28. Planet Neptune, *The Planets.*

1935

The Bowler, "Le Cricket" from *Sporting Sketches* (chor. Susan Salaman).

October 10. The Baron, *The Rape of the Lock* (chor. Andrée Howard).

April 7. Hercules, *The Descent of Hebe.*

1936

January 26. The Man She Must Marry, *Jardin aux Lilas.*

1937

January 7. Dancer, *Paleface,* A Floor Show, BBC-TV.

February 2. Dancer, *Hooey* and *Siesta.* BBC-TV.

February 19. Second Song, with Maude Lloyd, *Dark Elegies.*

March 2. Dancer, *Fugue for Four Cameras,* with Maude Lloyd, In "After Supper," a revue, *Fête d'Hébé* and *Wienerblut,* BBC-TV.

April 2. Dancer, *The Story of the Vienna Waltz,* BBC-TC.

April 10. The Squire, "Tommy and Sally," (chor. Andrée Howard) BBC-TV.

April 13. Dancer, *The Fairy Queen,* BBC-TV.

June 14. An Aristocrat in Love, *Gallant Assembly.*

June 29. Solo, "Captain Reginald D'Arcy of the Guards," *Boulter's Lock,* BBC-TV.

July 8. Dancer, *Relâche,* BBC-TV.

July 14. Dancer, *Romeo and Juliet,* with Maude Lloyd, BBC-TV.

July 24. Solo, *En Diligence,* BBC-TV.

October 25. *Full Moon.* "Waltz" and "Full Moon Blues" with Margaret Braithwaite, BBC-TV.

December 31. "La Musique Rejoint son Fauteuil," with Margot Fonteyn, BBC-TV.

1938
April 5. The Roué, *Wien,* BBC-TV.
April 24. Dancer, "Acis and Galatea" (chor. Andrée Howard), BBC-TV.
May 11. The Auctioneer, *The Emperor Jones.* BBC-TV.
May 29. The Scholar, "Master Peter's Puppet Show," BBC-TV.
June 15[?]. The Client, *Judgment of Paris.*
November 26. The Tirolese, with Maude Lloyd, *Soirée Musicale.*
December 5. Cavalier to the Italian Ballerina, *Gala Performance.*
December 13. Dancer, *Cinderella.* BBC-TV.

1939
February 3. *Soirée Musicale* and *Gala Performance.* The London Ballet, BBC-TV.

1940
January 11. The Dummy, tradition and the ordinary, *The Great American Goof* (chor. Eugene Loring).
February 1. Son Ami, *Quintet* (chor. Anton Dolin).
August 1. The Nobleman, *Goya Pastoral.*
[?]Pierrot. *Carnaval* (chor. Michel Fokine).

1941
November 12. King Bobiche, *Bluebeard* (chor. Michel Fokine).

1942
April 8. The Friend, *Pillar of Fire.*
September 8. The Father of Zemphira, *Aleko* (chor. Leonide Massine).
September 16. The Viceroy, *Dom Domingo de Don Blas* (chor. Leonide Massine).

1943
April. Tybalt, *Romeo and Juliet.*
October 20. He Wore a White Tie, *Dim Lustre.*

1951
February 28. The Father, *Lady of the Camellias.*
Tudor continued to perform various roles in his ballets until the late nineteen-fifties.

Glossary of Ballet Terms

Adagio The opening section of the conventional *pas de deux;* a slow series of movements.

Allegro Musical term denoting quick tempos; *petit allegri* are combinations of small movements that are executed at a quick pace.

Arabesque Tudor described the arabesque as a position where the body is supported on one leg while the other is extended in back with the arms harmoniously disposed.

Assemblé A jumping movement where the dancer throws one leg up and springs off the other: while ascending, the raised leg continues to rise. On landing both feet close down together.

Attitude The body is supported on one leg while the other leg is bent and is lifted either in front or behind. This position can be done turning.

Ballonnés The dancer springs, at the same time extending the bent leg loosely from the knee to front, side, or back and returning it on landing.

Baroque Dance A style of dance steps common to the seventeenth and eighteenth centuries where the dancers move in complex series of foot patterns while wearing heels and heavy period clothing.

Barre Ballet class begins with barre, in which the dancer lightly rests one hand on the barre, which surrounds the classroom.

Batterie General term for steps in which one leg beats against the other.

Bourrées Transfer of weight from one foot to the other while on pointe or demi-pointe. *Bourrées* usually travel across space.

Brisé A small travelling *assemblé* with a beat, ending on both feet, in which the dancer uses a spring to skim across the floor, usually sideways.

Cabrioles With one leg raised, the dancer springs from the supporting leg, which rises to beat beneath the raised leg. The dancer then lands on the supporting leg.

277

Cambré   A bend of the waist in any direction.

Chassé   The dancer slides a foot out in any direction, keeping the heel on the ground and bending that knee, and closes the second foot to it.

Changements   Starting in fifth position, the dancer springs up, straightening the knees in the air, and changes the feet before landing.

Corps de ballet   Group of supporting dancers usually moving in unison.

Coupé   A cutting step in which one foot is put down while picking up the other.

Czardas steps   Appeared in Hungarian ballrooms in mid–nineteenth century.

Diagonal   The dancer moves on the diagonal across the stage.

Downstage   The dancer moves downstage toward the audience.

Echappé   Both legs spring out simultaneously from bent to straight legs on to the toes or into the air and then return to the plié or bent-leg position. Good way to warm up on pointe.

Enchaînements   A phrase or pattern of steps.

Fifth position   A position in which the dancer stands with the heels of each foot touching the toes of the other.

Fish dive   A lift in which the dancer arches her back, lifts her head, and bends back her legs with her feet crossed.

Fouetté turns   A turn in which the dancer is supported on one leg while whipping the lifted leg from front to side while turning.

Galop   A lively round dance; its characteristic is a change of step, or hop, at the end of every phrase of music.

Glissade   A smooth travelling step gliding along the ground, beginning and ending with the feet together.

Grand battement   The dancer stands on one leg while flinging the other in the air, to the front, side, or back, and then bringing it back to the floor.

Grand jeté   A big jump in which one leg is thrown in front while the other is behind in what looks like a split in midair.

Grand jeté en tournant   The dancer stands with the right leg extended behind. Turning her back, she steps onto the right leg (now before her), and bends it, at the same time throwing her left leg past high into the air, and springs raising her arms to help herself. While in the air, the dancer turns to the front again, changing her legs so that she lands on the left leg in the position in which she began.

Jeté   A spring from one foot to the other.

Legato   Smooth and flowing steps.

Pas de Basque Associated with the Basque people. A step in three movements. The dancer sweeps the right foot to the front, then out to the side: next she springs onto it, and slides the left foot past to the front, when the right foot (now at the back) stretches and closes behind the left.

Pas de chat "Step of the cat"; the dancer throws a bent leg out to the side and brings the other one, also bent, immediately into the air underneath her while travelling sideways.

Pas de deux Duet, usually danced by a man and a woman.

Passé While supported on one leg, the dancer moves the lifted leg to the knee of the standing leg.

Piqué The dancer steps sharply onto one toe without bending the same knee. A turn on one leg in which the dancer spins round on one foot usually travelling across the stage.

Pirouette A turn on one leg usually with the raised foot touching the knee of the standing leg.

Plié A bending of the knee or knees.

Port de bras General term for movement and carriage of the arms.

Promenade The dancer turns steadily on the spot on one foot, moving the heel inch by inch.

Relevé The dancer raises the heel of the supporting foot off the ground, straightening the supporting knee.

Renversé The dancer sways her body from one side to the other as she turns, giving the impression that she is off balance.

Ronds de jambe à terre With the dancer standing on one leg, the other foot describes a semi-circle on the floor, moving steadily with the toe on the ground from front to side, side to back, and then past the stationary heel in a straight line to the front again.

Sauté A general term for "jump."

Temps de cuisse The dancer extends the right foot to the side and then closes it in fifth position as both knees bend, ready to spring sideways from the right leg, which lingers behind and closes just after the left.

Tours en l'Air The dancer springs straight up and turns while in the air. These are commonly begun with the feet in fifth position; the dancer springing up and turning in the direction of the front foot, with the feet held closely together.

Upstage The dancer moves upstage or away from the audience toward the back curtain.

Sources

Books

Anderson, Jack. *Choreography Observed.* Iowa City: University of Iowa Press, 1987.

Anthony, Gordon. *The Vic-Wells Ballet—Camera Studies.* London: George Routledge & Sons, 1938.

Balanchine, George, and Francis Mason. *The Complete Stories of the Great Ballets.* Garden City, New York: Doubleday and Company, 1977.

Ballet Rambert: 50 Years and On. Edited by Clement Crisp, Anya Sainsbury, Peter Williams (Scholars Press, 1976).

Barnes, Clive. *Inside American Ballet Theatre.* New York: Hawthorn Books, 1977.

Beaumont, Cyril W. *Complete Book of Ballets.* London: Putnam, 1937.

———. *Supplement to Complete Book of Ballets.* London: C. W. Beaumont, 1942.

———. *Ballets Past and Present.* London: Putnam, 1955.

———. *Dancers Under My Lens.* London: C. W. Beaumont, 1949.

———. *Design for the Ballet.* London: The Studio, 1937.

Belden, Kay. *The Story of Westminster Theatre.* London: Westminster Productions, 1965.

Blackmur, R. P. *Form and Value in Modern Poetry.* Garden City, New York: Doubleday Anchor Books, 1957.

Bradley, Lionel. *Sixteen Years of the Ballet Rambert.* London: Hinrichsen Edition, 1946.

———. Journals. Mss. in Gabrielle Enthoven Collection, Theatre Museum, London.

Brinson, Peter, and Clement Crisp. *Ballet for All.* London: Pan Books, Ltd., 1970.

Bristol, Caroline Jane. "Chronology of Tudor's Works, 1931–1954." Master's thesis, Juilliard School of Music, May 1956.

Bush, Ronald. *T. S. Eliot: A Study in Character and Style.* New York: Oxford University Press, 1983.

Chujoy, Anatole, and P. W. Manchester (ed.). *The Dance Encyclopedia,* revised and enlarged edition. New York: Simon & Schuster, 1967.

Clarke, Mary. *Dancers of Mercury.* London: Dance Books, 1962.

Cohen, Selma Jeanne (ed.). *Dance as a Theatre Art: Source Readings in Dance History from 1581 to the Present.* New York: Dodd, Mead, 1974.

Coton, A. V. *A Prejudice for Ballet.* London: Methuen, 1938.

———. *Writings on Dance, 1938–1968.* London: Dance Books, 1975.

Crisp, Clement (ed.), with Anya Sainsbury and Peter Williams. *Ballet Rambert: 50 Years and On.* London: Scholars Press, 1981.

Croce, Arlene. *AfterImages.* New York: Vintage Books, 1979.

Dance Notation Bureau, *Notated Theatrical Dances of Antony Tudor.* New York: Dance Notation Bureau, 1985.

De Mille, Agnes. *And Promenade Home.* New York: Da Capo Press, 1952.

———. *Dance to the Piper.* New York: Da Capo Press, 1958.

Denby, Edwin. *Looking at the Dance.* New York: Popular Library, 1943; reprint, Pelligrini and Cudahy, New York, 1948.

———. *Dance Writings.* New York: Alfred A. Knopf, 1986.

Deren, Maya. *The Divine Horsemen,* Foreword by Joseph Campbell. Reprint, New York: Delta, 1972.

De Valois, Ninette. *Come Dance with Me.* Cleveland, New York: World Publishing, 1957.

Dolin, Anton. *Ballet Go Round.* London: Michael Joseph, 1938.

Drew, David. *The Decca Book of Ballet.* London: Frederick Muller Ltd., 1958.

Franks, Arthur. *Twentieth Century Ballet.* New York: Pitman, 1954.

Garafola, Lynn. *Diaghilev's Ballets Russes.* New York: Oxford University Press, 1989.

Gruen, John. *The Private World of Ballet.* New York: Viking Press, 1975.

Hall, Fernau. *Ballet.* London: The Bodley Head, 1947.

———. "Dance Art and Beauty." N.D, NYPL Dance Collection.

Haskell, Arnold. *The Marie Rambert Ballet.* London: British-Continental Press, 1930.

——— (ed.). *Ballet Annual.* vols. 1–18. London: A. & C. Black, 1946–1963.

Heppenstall, Rayner. *Apology for Dancing.* London: Faber & Faber, Ltd., 1936.

Huxley, Michael. *Dance Analysis,* "History of a Dance." Janet Adshead (ed.). London: Dance Books, 1988.

Jaques-Dalcroze, Emile. *Rhythm, Music and Education,* London: Chatto & Windus, 1921.

Jowitt, Deborah. *The Dance in Mind.* Boston: David R. Godine, 1985.

Kersley, Leo, and Sinclair, Janet. *A Dictionary of Ballet Terms.* London: A. & C. Black, 1973.

Kirkland, Gelsey. *Dancing on My Grave.* New York: Jove Books, 1986.

———. *The Shape of Love.* New York: Doubleday & Co., 1990.

Koegler, Horst. *Oxford Dictionary of Ballet.* New York: Oxford University Press, 1977.

Leeper, Janet. *English Ballet.* London: King Penguin, 1944.

Markova, Alicia. *Markova Remembers.* London: Hamish Hamilton, 1986.

Massine, Leonide. *My Life in Ballet.* London: Macmillan, 1968.

Monahan, James. *The Nature of Ballet.* London: Pitman Publishing, 1976.

Nahum, Baron. *Baron at the Ballet.* New York: William Morrow, 1951.

Osato, Sono. *Distant Dances.* New York: Alfred A. Knopf, 1980.

Pask, Edward. *Ballet in Australia: The Second Act, 1940–1980.* Melbourne: Oxford University Press, 1982.

Payne, Charles. *American Ballet Theatre.* New York: Alfred A. Knopf, 1978.

Perlmutter, Donna. *Shadowplay: A Biography of Antony Tudor.* New York: Viking Books, 1991.

Rambert, Marie. *Quicksilver: An Autobiography.* London: MacMillan, 1972.

Robert, Grace. *The Borzoi Book of Ballets.* New York: Alfred A. Knopf, 1945.

Sawyer, Elizabeth. *Dance with the Music.* London: Cambridge University Press, 1985.

Siegel, Marcia B. *At the Vanishing Point.* New York: Saturday Review Press, 1972.

———. *Shapes of Change.* Boston: Houghton Mifflin, 1979.

Storey, Alan. *Arabesques.* London: Newman Wolsey, Ltd., 1948.

Van Praagh, Peggy. *How I Became a Ballet Dancer.* London: T. Nelson & Sons, 1959.

———, and Peter Brinson. *The Choreographic Art.* London: A. & C. Black, 1963.

Vaughan, David. *Frederick Ashton and His Ballets.* New York: Alfred H. Knopf, 1977.

Villella, Edward, with Larry Kaplan. *Prodigal Son: Dancing for Balanchine in a World of Pain and Magic.* New York: Simon & Schuster, 1992.

Sources

Walker, Kathrine Sorley. *De Basil's Ballets Russes.* New York: Atheneum, 1983.
Williamson, Audrey. *Ballet of Three Decades.* London: Salisbury Square, 1958.

Articles and Transcripts

Acocella, Joan. "Balancing Act." *Dance Magazine,* October 1990, pp. 55–57.
Allen, Zita. "The Heart of the Matter." *The Soho Weekly News,* 3 May 1979.
Anawalt, Sasha. *Ballet Review,* Winter 1988, pp. 15–16, 18, 19.
Anderson, Jack. "Antony Tudor Talks about His New Ballets." *Dance Magazine,* May 1966, pp. 42–45.
———. "The View from the House Opposite." *Ballet Review,* Vol. 4, no. 6, pp. 14–23.
Barlow, S. L. M. "Dim Lustre." *Modern Music,* November–December 1943, p. 55.
Barnes, Clive. "Antony Tudor: A Choreographer Far from Home." *Dance and Dancers,* July 1953, pp. 10–11.
———. "Jardin aux Lilas." *Dance and Dancers,* June 1959, pp. 18, 19, 28, 34.
———. "Tudor, the Seal of Greatness." *New York Times,* 23 January 1966.
Barret, Dorothy. "Understanding Antony Tudor." *Dance Magazine,* June 1945, pp. 12, 37, 49.
Beiswanger, George. "The Short Story Ballet." *Dance Magazine,* June 1942, pp. 10–11.
Bennahum-Chazin, Judith. "Shedding Light on Dark Elegies." *Society of Dance History Scholars Proceedings, North Carolina School of the Arts,* 1988, pp. 131–144.
———. "After 'Pillar of Fire'." *Choreography and Dance,* vol. 1, pt. 2, 1989, pp. 69–96.
———. "Scandinavian Memories of Antony Tudor." *Ballet Review* 20.4 Winter 1992–1993, pp. 46–51.
Bliss, Sally Brayley. "Personal Reminiscences." *Choreography and Dance,* vol. 1, pt. 2, 1989, pp. 27–37.
Brodie, Joan. "Shadowplay." *Dance Observer,* May 1948, p. 55.
Buckle, Richard. "Attitudes of Princely Contemplation." *The London Sunday Times,* 29 January 1967.
Butler, Gervaise. "Antony Tudor." *Dance Observer,* November 1943, p. 99.
Chapin, Isolde. "A Dressing Room Interview." *Dance Magazine.* January 1947, pp. 52–53.
Charlip, Remy. "Shadowplay." *The Village Voice,* 18 May 1967, p. 18.
Clarke, Mary. "Shadowplay." *The Dancing Times,* March 1967, pp. 290–293.
———. "Marie Rambert." *The Dancing Times,* July 1962, pp. 616–619.
Cohen, Selma Jeanne. "Time for Dance in Stockholm." *Saturday Review of Literature,* 26 June 1965, p. 55.
———. "Tudor and the Royal Ballet." *Saturday Review of Literature,* 13 May 1967, p. 74.
———. "Antony Tudor: Man Without a Theory." *Dance Magazine,* April 1954, pp. 15–16, 50–51.
———. "Antony Tudor: The Years in America and After." *Dance Perspectives,* vol. 18 (1963), pp. 1–104.
Chujoy, Anatole. "America Meets Antony Tudor." *Dance Magazine,* March 1940, pp. 18–19.
Como, William. "Celebrating 60 years." *Dance Magazine,* June 1987, pp. 100–120.
Coton, A. V. "Ballet Rambert." *Dance Magazine,* October 1949, pp. 8, 9, 40, 42.
Croce, Arlene. "Sweet Love Remembered." *The New Yorker,* 11 August 1975, p. 70.
"The Dance Magazine Awards." *Dance Magazine,* May 1974, pp. 38–42.
Dzhermolinska, Helen. "The Days of a Choreographer's Years." *The American Dancer,* June 1941, pp. 12–13.

Farkas, Mary. "Antony Tudor: The First Zen Institute." *Choreography and Dance,* vol. 1, pt. 2, 1989, pp. 59–67.

Frankenstein, Alfred. "Tudor's Definition of Ballet Is Not in Webster's Dictionary." *San Francisco Chronicle,* 6 February 1944.

Goodwin, Noel. "Lilac Garden." *Dance and Dancers,* January 1969, pp. 34–37.

Gowing, Lawrence. "Antony Tudor and the Dance Theatre." *Dancing Times,* July 1937, pp. 461–462.

Greenberg, Clement. "Dim Lustre." *The Nation,* 3 November 1945.

Gross, Michael. "The Turning Point." *New York Magazine,* June 17, 1991, p. 38.

Hall, Fernau. *Dance, Art, and Beauty,* 1947.

Haskell, Arnold. "Ballet in Britain, 1934–1944." *Dance Index,* October 1945, pp. 161–182.

Hill, Martha. "Antony Tudor and the Juilliard Years." *Choreography and Dance,* vol. 1, pt. 2, 1989, pp. 39–58.

Hobbs, Tarquin. "Born Under Mercury." *Ballet,* July 1946, pp. 57–59.

Horner, John. "Went to Ballet by Mistake and Became Balletmaster." *The Daily Worker* (London), 8 December 1938.

Howlett, Jasper. "The London Ballet." *The Dancing Times,* April 1939, pp. 19–20.

Hunt, Marilyn. "Maude Lloyd: In Conversation." *Ballet Review,* Fall 1983, pp. 5–26.

———. "Tudor on Tudor." *Dance Magazine,* May 1987, pp. 36–44.

Jordan, Stephanie. "Antony Tudor, His Use of Music and Movement." *Eddy Magazine,* Spring–Summer 1976, pp. 18–23.

Jowitt, Deborah. "Offenbach in the Underworld." *The Village Voice,* 20 October 1975.

———. "On Tudor's *Undertow.*" *The Village Voice,* 9 June 1992.

Kersley, Leo. "Choreographers of Today." *Ballet Today,* June 1960, pp. 13–15.

"Legend of Dick Whittington." *Oxford Companion to Children's Literature.* New York: Oxford University Press, 1984.

Lewis, Jean Battey. "Tudor, Subtle Analyst of Human Emotions," *Los Angeles Times,* 17 February 1974.

Lindamood, Peter. "With the Dancers." *Modern Music,* February 1940, p. 187.

Lloyd, Margaret. "Interview with Antony Tudor." *The Christian Science Monitor,* 24 February 1940.

———. "Romeo and Juliet." *The Christian Science Monitor,* 10 April 1943.

Lloyd, Maude. "Some Recollections of the English Ballet." *Dance Research,* vol. 3, no. 1, 1984, pp. 39–52.

MacInnes, Colin. "Backward Glances: Hugh Stevenson," *Dance and Dancers,* December 1960, pp. 9–11.

Manchester, P. W. "Ballet Rambert Makes Its Bow." *Theatre Arts,* July 1959, p. 66.

———. "Reflections after Reading *Shadowplay.*" *Ballet Review,* Spring 1992, pp. 74–77.

Martin, John. "The Year's Awards to Honor Antony Tudor" (editorial). *The New York Times,* 14 June 1942.

Maynard, Olga. "A Loving Memoir." *Dance Magazine,* August 1987, pp. 18–19.

Menuhin, Diana. "The Varying Moods of Tudor." *Dance and Dancers,* July 1955, pp. 9–10.

Monahan, James. "London Festival Ballet." *The Dancing Times,* June 1973, pp. 480–481.

Moore, Lillian. "Opera News Introduces Antony Tudor." *Opera News,* 27 November 1950, 4–5.

———. "Pillar of Fire." *The Dancing Times,* March 1966, p. 315.

Newman, Barbara. "Character and Caring." *Dance Magazine,* September 1987, pp. 54–59.

Ostlere, Hilary. "Deep Waters: Caught in the *Undertow.*" *Dance Magazine,* June 1992, pp. 52–55.

———. "The Tudor Spirit." *Ballet News,* April 1982, pp. 12–16.

Palatsky, Eugene. "Tudor and Others in Stockholm." *Dance and Dancers,* November 1963, p. 45.

Percival, John. "The Camargo Society." *Dance and Dancers,* January 1961, pp. 16–17.

———. "Antony Tudor: The Years in England." *Dance Perspectives,* Vol. 17 (1963), pp. 1–49.

———. "New Ballet of Surprise and Ideas." *The Times* (London), 26 January 1967.

———. "Talking about Antony Tudor's *Jardin aux Lilas.*" Sadler's Wells Theatre Community and Education Project, unpublished transcript, July 13, 1987.

———. "Working with Antony Tudor." Sadler's Wells Theatre Community and Education Project, unpublished transcript, 15 June 1987.

———. "Antony Tudor." *About the House,* Christmas 1966, pp. 32–35.

———. "Soirée Musicale." *Dance and Dancers,* April 1973, pp. 46–47.

Robertshaw, Ursula. "Rare Talent Returns to London." *The Illustrated London News,* 28 January 1967.

Rosen, Lillie F. "Talking with Antony Tudor." *Dance Scope,* vol. 9, no. 1, Fall 1974, pp. 14–24.

———. "Talking with Agnes de Mille." *Dance Scope,* vol. II, Fall/Winter 1976–77, pp. 8–17.

Sabin, Robert. "A Creative Crisis in Ballet Looms." *Dance Observer,* June 1945, p. 68.

———. "A Revolutionary Figure in Ballet Who Has Worked with Tradition." *Musical America,* February 1958, pp. 21, 71, 155, 304.

Schonberg, Harold. "Balanchine: Slaughtering a Sacred Cow." *Harper's Magazine,* April 1972, pp. 101–102.

Schulman, Jennie. "Antony Tudor in the Modern Ballet." *Dance Observer,* February 1951, pp. 20–21.

Schuman, William. "Toasting Tudor: The Capezio Awards." *Ballet Review,* Fall 1986, pp. 31–37.

"The Sitter Out." *The Dancing Times,* June 1939, p. 256.

Stevenson, Hugh. "Backward Glance." *Dance and Dancers,* December 1960, pp. 9–11.

Szmyd, Linda. "Antony Tudor: Ballet Theatre Years." *Choreography and Dance,* vol. 1, pt. 2, 1989, pp. 3–26.

"The Man Who Changed Ballet." *Dance and Dancers,* May 1987, pp. 14–17.

"Toasting Tudor: The Capezio Awards." *Ballet Review,* Fall 1986, pp. 31–37.

Todd, Arthur. "Approaches to Choreography." *Ballet Today,* London, pt. 1 (Oct. 1951), 16–18.

———. "Alcestis." *Dance Observer,* January 1961, pp. 7–8.

———. "Ballet at the Met." *Dance Observer* (March 1961), 42–43.

Topaz, Muriel. "Specifics of Style in the Works of Balanchine and Tudor." *Choreography and Dance,* vol. 1, pt. 1, 1988, pp. 2–36.

———. "Notating and Reconstructing for Antony Tudor." *Society of Dance History Scholars Proceedings,* February 1985, pp. 91–97.

Tudor, Antony. "America as the New Home of Ballet Tradition." *Musical Courier,* September 1944, p. 10.

———. "A Few Words Addressed to Dancers." New York Public Library Private Collection of Antony Tudor.

———. Interview by Marilyn Hunt, 22 August 1986. Transcription given to author by Marilyn Hunt.

———. "Movement in Opera." *Opera News,* 11 February 1961, pp. 8–13.

———. "Rambert Remembered." *Ballet Review,* Spring 1983, pp. 62–67.

———. "35th Anniversary Tribute to 'Mim.'" *Dance and Dancers,* October 1955, p. 12.

Turnbaugh, Douglas. "Good Guys vs. Bad Guys at Lincoln Center." *New York Magazine,* 29 May 1968, p. 51.

Van Praagh, Peggy. Interview by Margaret Dale, October 1978. Transcript, NYPL Dance Collection.

———. "Working with Antony Tudor." *Dance Research,* vol. II, no. 2, 1984, pp. 56–67.

Vaughan, David. "Antony Tudor's Early Ballets." *Society of Dance History Scholars Proceedings,* 1985, pp. 72–83.

———. "New York News Letter." *Dancing Times,* September 1975, pp. 644–645.

Walker, Kathrine Sorley. "The Choreography of Andrée Howard." *Dance Chronicle,* vol. 13, no. 3, 1990, pp. 265–358.

White, Franklin. "The Contribution of England to the Ballet." *Dance Observer,* February 1953, pp. 20–24.

Williams, Peter. "Shadowplay." *Dance and Dancers,* March 1967, pp. 10–18.

———. "The Making of a Muse: The Life and Achievements of Marie Rambert." *Dance Gazette* (London), no. 181, October 1982, pp. 2–5.

Windreich, Leland. "Shadowplay." *Ballet Review,* Spring 1992, pp. 77–80.

Interviews

Nora Kaye, Albuquerque, New Mexico, February 1985.

Muriel Topaz, New York City, February 1985, September 1989, and June 1991.

Nancy King Zeckendorf, New York City, conversations, 1987–1991.

Selma Jeanne Cohen, conversations, 1987–1991.

Diana Adams, New York City, telephone conversation, May 1987.

Maude Lloyd, London, April 1988.

Fernau Hall, London, April 1988.

Alicia Markova, London, April 1988.

Jane Pritchard, London, April 1988.

Nils Åke Haggböm, Stockholm, Sweden, June 1988.

Gerd Andersson, Stockholm, Sweden, June 1988.

Viola Aberle, Stockholm, Sweden, June 1988.

Lulli Svedin, Stockholm, Sweden, June 1988.

Mariane Orlando, Malmö, Sweden, June 1988.

Anne Borg, Oslo, Norway, June 1988.

Antony Tudor, New York and Los Angeles, telephone conversations and interviews, 1985 and 1986.

Antony Tudor, New York, interviewed by Marilyn Hunt, 22 August 1986.

Sallie Wilson, New York City, 1988, 1991, and June 1992.

Mary Farkas, New York City, January 1988.

Agnes de Mille, New York City, July 1989, and correspondence.

Bruce Marks, Boston, telephone interview, January 1991.

Donald Mahler, New York City, April 1991–1992.

Zachary Solov, New York City, telephone conversation, January 1991.

Anna Grete Stahl, New York City, June 1992.

Marilyn Hunt interviewed the following people:

—Ethan Brown, regarding *Pillar of Fire,* 19 January 1987.

—Hugh Laing, 9 May 1986. Oral History Project, NYPL Dance Collection.

—Kathleen Moore, regarding *Pillar of Fire,* 19 January 1987.

Notes

Chapter 1. Introduction to Tudor's Dominions: Studio and Stage

1. Isolde Chapin, "Dressing Room Interview," *Dance Magazine* (January 1947), 52–53.

2. Antony Tudor, "A Few Words Addressed to Dancers," undated, NYPL Dance Collection, Performing Arts Research Library.

3. Jennifer Predock-Linnell, interview with author (May 1991).

4. Selma Jeanne Cohen, "Antony Tudor," *Dance Perspectives,* 18 (1963): 76. See also Edwin Denby, "Tudor and Pantomime," *New York Herald Tribune,* 11 July 1943.

5. Lillie Rosen, "Talking with Antony Tudor," *Dance Scope,* vol. 9, no. 1 (Fall 1974): 19.

6. Malcolm Winter (10 November 1968), unidentified English newspaper clipping. NYPL Dance Collection.

7. Antony Tudor interview with Marilyn Hunt (22 August 1986).

8. Jennie Schulman, *Dance Observer* (June–July 1951), 93.

9. Dorothy Barret, "Understanding Antony Tudor," *Dance Magazine* (June 1945), 37.

10. Cohen, *Dance Perspectives,* 18 (1963): 84–85.

11. Muriel Topaz, interview with author (5 September 1989), New York City.

12. Muriel Topaz, "Notating and Reconstructing for Antony Tudor," *Society of Dance History Scholars Proceedings* (University of New Mexico, 1985), 96.

13. Alfred Frankenstein, *San Francisco Chronicle,* 1947 (no further details).

14. Muriel Topaz, interview with author (5 September 1989), New York City.

15. Elizabeth Sawyer, interview with author (4 September 1989), New York City.

16. Antony Tudor, interview with author (April 1986).

17. Elizabeth Sawyer, interview with author (4 September 1989).

18. Donald Mahler, letter to author (12 April 1991).

19. Gelsey Kirkland and Greg Lawrence, *The Shape of Love* (Doubleday & Co., 1990), 236. Antony Tudor, letter to Gelsey Kirkland.

20. *Dance Perspectives,* 18 (1963), 79.

21. Donald Mahler, letter to author (12 April 1991).

22. Sallie Wilson, interview with author (December 1991), New York City.

23. Rayner Heppenstall, *Apology for Dancing* (London: Faber & Faber Ltd., 1936), 222.

24. Muriel Topaz, interview with author (5 September 1989), New York City.

25. Bruce Marks, telephone interview with author (January 1992).

26. Alan Storey, "Three Men," in *Arabesques* (London: Newman Wolsey Ltd., 1948), 136.

27. Jennie Schulman, *Dance Observer* (February 1951), 21.

28. Jean Battey Lewis, "Tudor, Subtle Analyst of Human Emotions," *Los Angeles Times,* 17 February 1974.

29. Zita Allen, "The Heart of the Matter," *SoHo Weekly News* (3 May 1979).
30. Ibid.
31. Lewis, "Tudor, Subtle Analyst."
32. Zita Allen, *Soho Weekly News* (3 May 1974).
33. John Percival, *About the House* (Christmas 1966), 34.
34. R. P. Blackmur, *Form and Value in Modern Poetry* (Garden City, New York: Doubleday Anchor Books, 1957), 339.
35. Stephanie Jordan, "Antony Tudor: His Use of Music and Movement," *Eddy Magazine* (Spring–Summer 1976), 19.
36. Edwin Denby, *Looking at the Dance* (New York: Popular Library, 1943), 361.
37. Agnes de Mille, interview with the author, July 1989, New York City.

Chapter 2. Early Years: Childhood in London and the Discovery of a Passion
1. Antony Tudor, interview with author, April 1986.
2. Antony Tudor, interview with author, April 1986.
3. Robert Sabin, "A Revolutionary Figure in Ballet Who Has Worked with Tradition," *Musical America* (February 1958), 21.
4. Ibid.
5. Antony Tudor, interview with author, April 1986.
6. In 1912, Dalcroze wrote a fascinating essay, "How to Revive Dancing," which described the mission of early modern dance. He defined dancing as "the art of expressing emotion by means of body movements." See Emile Jaques-Dalcroze, *Rhythm, Music and Education* (London: Chatto & Windus, 1921), 176.
7. Lynn Garafola, *Diaghilev's Ballets Russes* (New York: Oxford University Press, 1989), 336.
8. Ibid., 365.
9. Ibid., 330.
10. Agnes de Mille, *Dance to the Piper* (New York: Da Capo Press, 1958), 182.
11. John Percival, "The Camargo Society," *Dance and Dancers* (January 1961), 16.
12. In her autobiography, *Quicksilver,* Rambert says that Tudor first visited with her in 1929. Tudor was probably correct in saying it was 1928 because he recalled that the following summer, 1929, the Diaghilev Ballets Russes gave its last season in London, leaving the city desolate of theater dance.
13. Marie Rambert, *Quicksilver* (London: MacMillan, 1972), 145.
14. Antony Tudor, interview with author (April 1986). Tudor told the story of how his mother attended a showing of his *Pillar of Fire:* When he asked how she liked the ballet, she responded in tears that she did not know which of the ballets was choreographed by him but that she was overjoyed to be the mother of Antony Tudor.
15. Antony Tudor, interview with author, April 1986.
16. Sabin, "Revolutionary Figure," 71.
17. Antony Tudor, interview with author, April 1986.
18. Antony Tudor, "Rambert Remembered," *Ballet Review* (Spring 1983), 62.
19. Ibid., 63.
20. Ibid., 64.
21. John Percival, "Antony Tudor," *About the House* (Christmas 1966), 33.
22. Fernau Hall, interview with author, April 1988.
23. Rambert, *Quicksilver,* 45.

24. Maude Lloyd, interview with author, April 1988.

25. Tarquin Hobbs, "Born Under Mercury," *Ballet* (July 1946), 58.

26. Maude Lloyd, interview with author, April 1988.

27. Helen Dzhermolinska, "The Days of a Choreographer's Years," *American Dancer* (June 1941), 13.

28. Rambert, *Quicksilver*, 145.

29. Elisabeth Schooling, letter to author (8 January 1989).

30. Lionel Bradley, *Sixteen Years of the Ballet Rambert* (London: Hinrichsen Edition Limited, 1946), 20–21.

31. "The Sitter Out," probably P. J. S. Richardson, *The Dancing Times* (December 1931), p. 232.

32. Cyril Beaumont, *Dances Under My Lens* (London: C. W. Beaumont, 1949), 33.

33. Ibid.

34. Ibid.

35. Fernau Hall, interview with author, April 1988.

36. Lionel Bradley, *Journals* (February 7, 1938), MS. in Gabrielle Enthoven Collection, Theatre Museum in London.

37. Tudor's fascination with the theater did not diminish while he worked with Rambert. Vivienne Browning received a letter from Antony Tudor about Gerald Godley, a mutual friend. Subsequently she wrote *The Dancing Times* (November 1990, p. 141):

> Shortly after he left school, Antony Tudor joined a little amateur dramatic society attached to a church in the City Road about 7 minutes from the Angel crossroads. It was very small and they rehearsed in the church parish hall which lay at the back of the church. It had a platform at one side which could be rigged up into some distant imitation of a stage. [In his letter to Browning, Tudor described his adventure:] "You can't imagine how lowly this was when I tell you that my first role there was Professor Higgins in *Pygmalion* and I think my Eliza was named Dora Cannon. . . . We were supervised by a lady coach from the LCC [London County Council], and as we were few in number, she had to bring students of hers from other places when it became necessary for a production. This happened soon because it was decided to do *Twelfth Night*. With such small choice available . . . it fell upon me to be cast as Malvolio. I must have been all of 18 years of age. . . ." The following year a different lady was sent to take charge and I played my third lead in something from French's catalogue named *Summer Lightning*. I was then introduced by the new directrice to the City Literary Institute near Southampton Row where a lot of time was spent rehearsing scenes from the ancient classics and there my favorite partner was named Jeanne Laskey. There was also a teacher who gave dancing classes, who also taught at the St Pancras People's Theatre. I was introduced there to dance in their annual panto [pantomime], partnering a girl in the Grieg's Dance in the Hall of the Mountain King. This was the first time I met Gerald Godley. . . . By this time I was more into dance and had become acquainted with Rambert, who was the very first to point out to me that I had a noticeable cockney accent. Then she was getting near to opening the Ballet Club and it offered me a job which I accepted with the proviso that I would be allowed to choreograph a ballet and that it should be based on *Twelfth Night* since I was au courant with the subject and sure enough soon after this, I presented my first ballet, *Cross-Garter'd.*"

Another theater that encouraged experimental plays where someone of Tudor's dramatic sensibility could learn was the Westminster. In addition to the remarkable plays commissioned by Ashley Dukes to be performed at the Mercury, the Westminster Theatre provided another London venue for more experimental plays. In 1931, Anmer Hall created the Westminster Theatre, where Tudor was a regular visitor. Ivan Turgenev, Eugene O'Neill, Jean Jacques Bernard and Henrik Ibsen were some of the well represented playwrights. See K. D. Belden, *The Story of Westminster Theatre,* London: Westminster Productions, 1965. Audiences came to reassess the old and respond to the new. They were a select group genuinely interested in discovery. But the support that the Dukes received, especially for ballet, had been nurtured by the tremendous impact of the Ballets Russes upon earlier London spectators.

38. John Percival "Talking about Antony Tudor's *Jardin aux Lilas* (July 13, 1987). Sadler's Wells Theatre Community and Education Project, unpublished transcript, 13 July 1987.

39. Marilyn Hunt, *Ballet Review* (Fall 1983), 14.

40. One wonders at this remark, as not many speak of Rubenstein as a great choreographer.

41. David Vaughan, *Frederick Ashton and His Ballets* (New York: Alfred Knopf, 1977), 120.

42. Kathrine Sorley Walker, "The Choreography of Andrée Howard," *Dance Chronicle,* vol. 13, no. 3 (Summer 1991), 270.

43. Alicia Markova, interview with the author, London, April 1988.

44. Rambert, *Quicksilver,* 151.

45. Hugh Laing, interview with Marilyn Hunt, 9 May 1986. Oral History Project, NYPL Dance Collection.

46. Ibid.

47. Elisabeth Schooling in letter to the author, January 8, 1989.

48. De Mille, *Dance to the Piper,* 80.

49. *New Clarion,* December 17, 1932.

50. *Christian Science Monitor,* December 31, 1932.

51. *Eastern Daily Press,* Norwich, England, December 6, 1932.

52. Arnold Haskell, *New English Weekly,* December 15, 1932.

53. *Time and Tide,* December 10, 1932.

54. Unidentified newspaper clipping, NYPL Dance Collection.

55. *New English Weekly,* December 15, 1932.

56. Unidentified newspaper clipping, Rambert Archives.

57. Constant Lambert, *Sunday Referee,* May 14, 1933.

58. Ibid.

59. Ibid.

60. Rambert, *Quicksilver,* 151.

61. Mary Clarke, *The Dancers of Mercury* (London: Adam & Charles Black, 1962), 86.

62. Ibid.

63. Ibid., 92.

64. Baron Nahum, *Baron at the Ballet* (New York: William Morrow, 1951), 51.

65. Ronald Bush, *T. S. Eliot: A Study in Character and Style* (New York: Oxford University Press, 1983), 123 and 169.

66. Ibid., 205.

67. *T. S. Eliot, Collected Poems* (New York: Harcourt Brace and Company, 1930), 183.

68. "The Legend of Dick Whittington," *Oxford Companion to Children's Literature* (New York: Oxford University Press, 1984), 149.

69. Colin MacInnes "Backward Glances: Hugh Stevenson," *Dance and Dancers* (December 1960), 9.

70. Rambert, *Quicksilver,* 150.

71. Percival disagreed in his article about *The Planets* in *Dance and Dancers* (February 1985), 28.

72. Rambert, *Quicksilver,* 151.

73. Marilyn Hunt, *Ballet Review* (Fall 1983), 12.

74. Percival, *Dance Perspectives,* no. 17 (1963), 22.

75. Percival, *Dance and Dancers* (February 1985), 20.

76. Young's copious notes are located in the Rambert Archives in London.

77. Marilyn Hunt, *Ballet Review* (Fall 1983), 13.

78. Percival *Dance Perspectives,* no. 17 (1963), 22.

79. Clarke, *Dancers of Mercury,* 96.

80. Haskell quoted in ibid., 96.

81. Parts of *The Planets* may well be performed once again. In 1993, Sally Brayley Bliss, executrix of the Tudor estate, and Muriel Topaz, notator, went to London to work with Elisabeth Schooling and Maude Lloyd, who were able to recall a great deal. They hope to reconstruct "Neptune," "Venus," and "Mars."

82. Tudor, interview with author, April 1906.

83. Dzhermolinska, "Days of a Choreographer's Years," 20.

84. Peggy van Praagh interviewed by Margaret Dale, New York Public Library transcription, October 1970.

85. Fernau Hall, *Dancing TImes* (June 1937), 293.

86. Kathrine Sorley Walker, "The Choreography of Andrée Howard," *Dance Chronicle,* vol. 13, no 3 (1990–91), 279.

87. Clarke, *Dancers of Mercury,* 100.

88. Hunt, *Ballet Review* (Fall 1983), 12.

89. Janet Leeper quoted in Hilary Ostlere, "The Tudor Spirit," *Ballet News* (April 1992), 14.

90. Marilyn Hunt, *Ballet Review* (Fall 1983), 12.

91. Elisabeth Schooling, letter to author, 4 December 1988.

92. De Mille, *Dance to the Piper,* 194.

93. Fernau Hall, interviewed by author, April 1990.

94. Ibid.

95. R.W.S., *Catholic Herald,* Manchester, 31 January 1936.

96. Hugh Laing, interviewed by Marilyn Hunt, May 9, 1961.

97. Beaumont, *Daily Telegraph,* 9 February 1935.

98. Arnold Haskell, *New English Weekly,* 23 May 1936.

99. *The Times,* 9 February 1937.

100. Heppenstall, *Apology for Dancing,* 220.

101. *Journals,* Lionel Bradley, 6 December 1936, Theatre Museum in London.

102. Elisabeth Schooling, letter to author, December 4, 1988.

103. When I interviewed Markova in London, she did not remember working with Tudor on this production. She believed that Robert Helpmann did the choreography and that Tudor had created another version.

Chapter 3. Revelation of a Major Talent
1. Cyril Beaumont, *Ballets, Past and Present* (London: Putnam, 1955).
2. Hugh Laing, interview with Marilyn Hunt, May 9, 1986.
3. Percival, *Dance Perspectives,* 17, p. 22.
4. Laing, interview by Marilyn Hunt, May 9, 1986.
5. Edward VII, King of England from 1901–10, son of Queen Victoria.
6. De Mille, *Dance to the Piper,* 65.
7. Fernau Hall mentioned that in 1985, Tudor added a special light from the house that poured into the garden so that the scene had a more realistic effect.
8. Bradley, *Journals,* February 15, 1937.
9. Percival, *Dance Perspectives,* no. 17 (1963), 25.
10. Jordan, "Antony Tudor," 19.
11. Lionel Bradley, *Journals* (July 19, 1938).
12. Jordan, "Antony Tudor," 20.
13. Antony Tudor, interview with the Author, April 1986, New York City.
14. Percival, "Working with Antony Tudor," Sadler's Wells Theatre Community and Education Project, June 15, 1987.
15. Jack Anderson's quote from "View from the House," *Ballet Review,* vol. 4, no. 6 (1974), 18.
16. Peggy van Praagh's quote from "Working with Antony Tudor," *Dance Research Journal,* vol. 2, no. 2, p. 58.
17. Marilyn Hunt, *Ballet Review* (Fall 1983), 11.
18. Maude Lloyd in interview with author, April 1988, London.
19. Clive Barnes, "Jardin aux Lilas," *Dance and Dancers* (June 1959), 18.
20. Lillie Rosen, *Dance Scope,* vol. 9, no. 1 (Fall 1974–75), 15.
21. Elizabeth Sawyer, interview with author, September 1989.
22. Noel Goodwin, "Lilac Garden," *Dance and Dancers,* January 1969, 37.
23. Hugh Laing interview with Marilyn Hunt, May 1986.
24. Maude Lloyd, quoted in John Percival, "Working with Antony Tudor," Sadler's Wells Theatre Community and Education Project, June 15, 1987.
25. Rayner Heppenstall, *The New English Weekly* (February 13, 1936), 353.
26. Arnold Haskell, *Daily Telegraph,* 27 January 1936.
27. Menuhin was married to the Rambert dancer Diana Gould.
28. Arnold Haskell, *New Statesman,* 1 February 1936.
29. Bradley's *Journals* are a wonderful source for detailed descriptions of Tudor's ballet costumes, especially those of the premiere London Ballet performances.
30. Hugh Laing, interview with Marilyn Hunt, May 9, 1986.
31. Antony Tudor, *Ballet Review* (Fall 1986), 34.
32. Laing interview with Marilyn Hunt, May 9, 1986.
33. See also Janet Adshead, *Dance Analysis, Theory and Practice* (London: Dance Books, 1983) for an excellent discussion of *Dark Elegies* by Michael Huxley.
34. Reissued on a Preiser Label. George Dorris, letter to author, March 8, 1988.
35. English translation by Walter Breen, 1971, from the disc "Songs of Gustav Mahler," Parnassus 4 label, recorded in 1971. Only five of over three hundred poems were set by Mahler, whose "Kindertotenlieder" was first performed in 1905 under his own direction, at a concert sponsored by the Vereinigung Schaffender Tonkunstler.
36. Peggy van Praagh interview with Margaret Dale, New York Public Library, October 1978.

37. Margaret Lloyd, *Christian Science Monitor,* February 24, 1940.

38. They printed the fact that *Dark Elegies* had to be postponed from February 15 to February 19; *Morning Advertiser,* February 15, 1937.

39. Fernau Hall, *The Dancing Times* (August 1937), 556.

40. A. V. Coton, *Writings on Dance* (London: Dance Books, 1938), 62.

41. *The News Chronicle,* 22 February 1937.

42. Mr. Forsyth of *The Star,* February 30, 1937.

43. *Daily Telegraph,* February 20, 1937.

44. Peggy van Praagh, "Working with Antony Tudor," *Dance Research* (Summer 1984), 57.

45. De Mille, *Dance to the Piper,* 194.

46. Van Praagh, "Working with Antony Tudor," 65.

47. Ibid., 66.

48. Rosen, "Talking with Antony Tudor," 17.

49. Peggy van Praagh, and Peter Brinson, *The Choreographic Art* (London: A & C Black, 1963), 73.

50. Van Praagh, "Working with Antony Tudor," 57.

51. David Vaughan, "Antony Tudor's Early Ballets," *Society of Dance History Scholars Proceedings* (University of New Mexico, 1985), 78.

Chapter 4. Tudor on His Own

1. Hilary Ostlere, "The Tudor Spirit," *Ballet News,* April 1982, p. 14.

2. For details, see Appendix C.

3. Dallas Bower, letter to author (7 December 1990).

4. "Working with Tudor," Sadler's Wells Theatre Community and Education Project (15 June 1987).

5. Van Praagh, "Working with Antony Tudor," 60.

6. Agnes de Mille interview with author, July 1989.

7. De Mille, *Dance to the Piper,* 193.

8. Van Praagh, "Working with Antony Tudor," 60.

9. John Irvine, *Isis* (16 June 1937).

10. Horace Horsnell, *Oxford Mail* (15 June 1937).

11. Irvine, *Isis.*

12. A. V. Coton, *Writings on Dance, 1938–1968* (London: Dance Books, 1975), 68.

13. Illustrated in Cyril Beaumont's *Design for the Ballet* (London: The Studio, Ltd., 1938).

14. Lionel Bradley, *Journals* (5 December 1938).

15. Lawrence Gowing, "Antony Tudor and the Dance Theatre," *Dancing Times,* London, July 1937, p. 462.

16. Lionel Bradley, *Journals* (5 December 1938).

17. Horace Horsnell, *The Observer,* 19 June 1938.

18. See *Dance Perspectives,* Chronology no. 18 (1963).

19. Vaughan, *Frederick Ashton and His Ballets,* 165.

20. David Drew, *The Decca Book of Ballet* (London: Frederick Muller Ltd., 1958), 390.

21. Grace Robert, *Borzoi Book of Ballet* (New York: Alfred Knopf, 1945), 198.

22. Lionel Bradley, *Journals* (5 December 1938).

23. F. T. *Daily Telegraph,* 24 January 1939.

24. Antony Tudor, interview with Marilyn Hunt (22 August 1986).

25. *The [London] Times,* 28 November 1938.

26. Coton, *Writings on Dance,* 68.

27. Lionel Bradley, *Journals* (9 January 1939).

28. "The Sitter Out," *The Dancing Times,* June 1939, p. 256.

29. John Percival, *Dance and Dancers* (April 1973), 46.

30. Maude Lloyd, interview with author (14 April 1988).

31. *Weekly Illustrated* (31 December 1938), 20.

32. Donna Perlmutter, *Shadowplay: A Biography of Antony Tudor* (New York: Viking, 1991), 98.

33. *London Observer* (6 March 1938).

34. Jasper Howlett, "The London Ballet," *The Dancing Times* (April 1939), 19.

35. *A News Chronicle* (24 November 1938).

36. See A. V. Coton about the London Ballet's title and why Tudor thought it pretentious.

37. Coton, *Writings on Dance,* 65.

38. Lionel Bradley, *Journals,* 5 December 1938.

39. Janet Sinclair, *Dance and Dancers* (January 1989), 18–20.

40. Francis Mason's and George Balanchine's *Complete Stories of the Great Ballets* (Garden City, N.Y.: Doubleday and Company) 1977.

41. Fernau Hall, *Carnaval* (March/April 1947), 109.

42. *New English Weekly* (13 February 1936), 353.

43. Tudor in interview with author, April 1986.

44. De Mille accused Tudor of taking "a lot of the Degas for his *Gala Performance.*" Lillie Rosen, "Talking with Agnes de Mille," *Dance Scope,* vol. 11 (Fall/Winter 1976/77), 12.

45. Films of several productions of *Gala Performance* may be found in the New York Public Library.

46. Maude Lloyd in interview with author, April 1988.

47. Lionel Bradley, *Journals* (5 December 1938; 9 January 1939).

48. Cyril Beaumont, *International Post* (6 April 1939).

49. See appendix for a list of his works.

50. Tudor and Laing may also have left England because they secretly hungered for a new, more open place to start their life. Until recently, the legal approach to homosexuality in England was harsh and punitive. They may have hoped for a more enlightened attitude in New York.

Chapter 5. To America: The Journey That Lasts a Lifetime

1. See Linda Szmyd's article in *Choreography and Dance,* vol. 1, part 2 (1989), 3.

2. Vaughan, *Frederick Ashton and His Ballets,* 177.

3. Letters in Tudor's personal file, Dance Collection at the New York Public Library.

4. John Martin, *New York Times,* 8 October 1939.

5. Charles Payne, *American Ballet Theatre* (New York: Alfred A. Knopf, 1978), 41.

6. Payne, *American Ballet Theatre,* 41.

7. *New York Times,* 21 July 1940.

8. John Martin, *New York Times,* 26 July 1941.

9. *New York Sun,* 26 February 1941.

10. Peter Lindamood, *Modern Music* (February 1940), 187.

11. Anatole Chujoy, "America Meets Antony Tudor," *Dance Magazine* (March 1940), 18.

12. Neil Weilbel, *Chicago Music News,* 21 March 1940.

13. R. Pleasant, *New York Herald Tribune,* 22 March 1940.

14. At the time that Tudor left London, van Praagh and Lloyd arranged for the London Ballet to perform for Harold Rubin, a wealthy entrepreneur who created a new organization, the Arts Theatre Club in Leicester Square.

15. Van Praagh, "Working with Antony Tudor," *Dance Research Journal,* vol. 2, no. 2, 1984, p. 62.

16. Rambert, *Quicksilver,* 171.

17. Pitts Sanborn, *World Telegram* N.D. newspaper clipping.

18. John Martin, *New York Times,* 2 August 1940.

19. Walter Terry, *New York Herald Tribune,* 2 August 1940.

20. Claudia Cassidy, *Chicago Journal of Commerce* (December 1940).

21. Helen Dzhermolinska, *American Dancer* (June 1941), 28.

22. Selma Jeanne Cohen, *Dance Perspectives,* no. 18 (1963), 54.

23. Solov interview with author, January 15, 1991.

24. *New York Times,* 14 January 1948.

25. *The Archives for the Performing Arts Quarterly,* San Francisco vol. 3, no. 4 (1986/87), 17.

26. *New York Herald Tribune,* 14 January 1948.

27. *The Christian Science Monitor,* February 21, 1942.

28. Agnes de Mille, *Dance to the Piper* p. 194.

29. *The Dancing Times* (March 1942), 294.

30. *The New York Times,* 14 June 1942.

31. Barbara Newman, "Character and Caring," *Dance Magazine* (September 1987), 54.

32. Newman, *"Character and Caring,"* 54.

33. Ibid., 56.

34. Undated letter from Tudor to Maina Gielgud of the Australian Ballet, Dance Collection at the NYPL for the Performing Arts.

35. Antony Tudor, interview with Marilyn Hunt (22 August 1986.)

36. Ibid.

37. Charles Payne, *American Ballet Theatre,* 118.

38. Ostlere, *Ballet News* (April 1982), 14.

39. See Marcia Siegel's wonderfully detailed description of *Pillar of Fire* in *Shapes of Change* (Boston: Houghton Mifflin, 1979), 152–163.

40. Edwin Denby, *Dance Writings* (New York: Alfred A. Knopf, 1986), 262.

41. Edwin Denby, *New York Herald Tribune,* 11 July 1943.

42. Jack Anderson, *Choreography Observed* (Iowa City: University of Iowa Press, 1987), 96.

43. *New York Times,* 22 April 1942.

44. Interview with Anna Kisselgoff, *New York Times,* 26 April 1942.

45. Siegel, *Shapes of Change,* 162.

46. Ibid.

47. *New York Times,* 26 April 1942.

48. Sallie Wilson, interview with author, January 1988.

49. Sono Osato, *Distant Dances* (New York: Alfred A. Knopf, 1980), 205.

50. John Gruen, *The Private World of Ballet* (New York: Viking Press, 1973), 261.

51. No date for this article by Frankenstein in the *San Francisco Chronicle.*

52. Alicia Markova, *Markova Remembers* (London: Hamish Hamilton, 1986), 108.

53. Frankenstein, op. cit.

54. S. L. M. Barlow, *Modern Music* (Nov.–Dec. 1943), 55.

55. Denby, *Looking at the Dance,* 230.

56. Osato, *Distant Dances,* 200.

57. Ibid., 203.

58. Ibid.

59. Rosen, *Dance Scope,* vol. 11, no. 1 (1976/77), 12. De Mille suggested that for Tudor's ballet, "the whole ballroom scene was taken from my film of *Romeo and Juliet* with Norma Shearer."

60. Osato, *Distant Dances* 201.

61. Jowitt, *The Dance in Mind* (New York: David R. Godine, 1985), 198.

62. Osato, *Distant Dances,* 202.

63. Denby, *Looking at the Dance,* 220.

64. *New York Times,* 18 April 1943.

65. *New York Sun,* 7 April 1943.

66. *New York Herald Tribune,* 23 October 1943.

67. Marcia Siegel, *At the Vanishing Point* (New York: Saturday Review Press, 1972), 43.

68. The following remarks by Tudor were taken from the pages of "Dim Lustre: Another Story," in the Dance Collection, New York Public Library.

69. Jennie Schulman, *Dance Observer* (February 1951), 21.

70. I saw a film in the New York Public Library from the 1940's with Nora Kaye and Hugh Laing.

71. *New York Times,* 12 May 1985.

72. Ibid.

73. *New York Herald Tribune,* 21 October 1943.

74. *New York Times,* 21 October 1943.

75. *Dance Observer* (November 1943), 99.

76. Maude Lloyd, interview with author, April 1988.

77. Sallie Wilson, interview with author, January 1988.

78. Antony Tudor, "America as the New Home for Ballet Tradition," *Musical Courier,* (September 1944), 10.

79. Szmyd, *Choreography and Dance,* 9.

80. Szmyd, *Choreography and Dance,* 11.

81. Charles Payne, *American Ballet Theatre,* 91.

82. Fernau Hall, interview with author, April 1988.

83. Dorothy Barret, "Understanding Antony Tudor," *Dance Magazine* (June 1945), 12.

84. Isolde Chapin, "Dressing Room Interview," *Dance Magazine* (January 1947), 52–53.

85. Fernau Hall, interview with author, April 1988.

86. Selma Jeanne Cohen, *Dance Perspectives,* no. 18 (1963), 85.

87. Ibid.

88. *Ballet Review* (Fall 1986), 36.

89. Ibid.

90. Edwin Denby, *New York Herald Tribune,* 15 April 1945.

91. Fernau Hall, *Dance, Art and Beauty,* n.d.

92. Tudor, interview by Marilyn Hunt, New York (22 August 1986).

93. Quote from "Caught in the Undertow," *Dance Magazine* (June 1992), 54.

94. Gruen, *Private World of Ballet,* 262.

95. Fernau Hall, interview with author, April 1988.
96. *Ballet Review,* vol. 4, no. 6 (1974), 20.
97. Edwin Denby, *New York Herald Tribune,* 15 April 1945.
98. Deborah Jowitt, "Primal Storms," *Village Voice,* 9 June 1992.

Chapter 6. Disenchantment in New York
1. *New York Post,* 23 November 1945.
2. *Daily Mirror,* 23 November 1945.
3. Gregrory Carmichael, "The Dancing Tudor," *Tricolor Magazine* (May 1945), 79.
4. Antony Tudor, interview with author, April 1986, New York. In retrospect, one may sympathize with the English attitude that shunned Tudor. De Valois's book, *Come Dance with Me,* which covers the 'thirties and 'forties of her career, makes no mention of Tudor's name. Photos illustrate the importance of dancers dressed in military uniforms during World War II.
5. Selma Jeanne Cohen, and Al Pischl, "The American Ballet Theatre," *Dance Perspectives,* no. 6 (1960), 51.
6. Ibid.
7. Isolde Chapin, "Dressing Room Interview," *Dance Magazine* (January 1947), 52.
8. Ostlere, "Tudor Spirit," 16.
9. George Balanchine's and Francis Mason's *The Complete Stories of Great Ballets* 1977 (New York: Doubleday), 571.
10. Ibid.
11. Zachery Solov in telephone conversation with author, 10 January 1991.
12. Robert Sabin, *Musical America* (May 1948), 272.
13. Joan Brodie, *Dance Observer* (May 1948), 55.
14. *New York Herald Tribune,* 15 April 1948.
15. *New York Times,* 15 April 1948.
16. *Morning Telegraph,* 16 April 1948.
17. *New York Herald Tribune,* 18 April 1948.
18. Fernau Hall, interview with author, April 1988, in London.
19. Ibid.
20. Agnes de Mille, interview with author, July 18, 1989, in New York.
21. Cohen and Pischl, *Dance Perspectives,* no. 6 (1960), 61.
22. Diana Adams, phone conversation with author, May 1987, New York.
23. *New York Herald Tribune,* 16 July 1949.
24. *Springfield Massachusetts Republican,* 16 July 1949.
25. *Springfield Massachusetts Republican,* 17 July 1949.
26. Schulman, *Dance Observer* (February 1951), 21.
27. Antony Tudor, interview with Marilyn Hunt, August 22, 1986, New York.
28. *Dance Observer* (February 1951), 21.
29. *New York Herald Tribune,* 4 May 1950.
30. *New York Times,* 4 May 1950.

Chapter 7. Tudor the Educator
1. His last ballet at Jacob's Pillow was *Little Improvisations* in 1953.
2. Films being unclear, and no notation.
3. 27 November 1950.
4. *Dance Perspectives,* no. 18 (1963), p. 62.

5. Lillian Moore "New York Notes: A New Tudor Ballet," *The Dancing Times* (April 1951), 397.

6. Ibid., 398.

7. Ibid.

8. *New York Herald Tribune,* 1 March 1951.

9. *New York Times,* 1 March 1951.

10. Jennie Schulman, *Dance Observer* (June–July 1951), 93.

11. Moore, *The Dancing Times* (June 1951), 519.

12. Selma Jeanne Cohen, *Dance Perspectives,* no. 18, p. 60.

13. *New York Herald Tribune,* 19 August 1951.

14. Ibid.

15. Sallie Wilson in an interview with the author in New York, January 6, 1988.

16. *Dance Perspectives,* no. 18, p. 68.

17. *New York Times,* 1 December 1951.

18. *New York Herald Tribune,* 2 December 1951.

19. Cyril Beaumont, *Ballets Past and Present,* 173.

20. *Dance Perspectives,* no. 18, 63.

21. John Gruen, *The Private World of Ballet,* 261.

22. Elizabeth Sawyer in interview with the author, September 4, 1989, New York City.

23. *New York Times,* 27 February 1952.

24. Sallie Wilson in interview with the author, January 6, 1988, New York City.

25. Doris Hering, *Dance Magazine* (May 1952), 45.

26. *New York Herald Tribune,* 27 February 1952.

27. Janet Rowson Davis, letter to author, December 15, 1991.

28. Caroline Bristol Britting in letter to author, January 15, 1991.

29. *Dance Perspectives,* no. 18, p. 67.

30. *New York Herald Tribune,* 29 August 1953.

31. *Springfield Republican,* 19 August 1953.

32. Nils Åke Haggböm in conversation with the author, June 1988, Stockholm.

33. *Springfield Republican,* 30 August 1953.

34. Martha Hill, "Antony Tudor and the Juilliard Years," *Choreography and Dance,* vol. 1, pt. 2 (1989), 50.

35. *Dance Perspectives,* no. 18, p. 67.

36. Martha Hill, *Choreography and Dance,* vol. 1, pt. 2 (1989), 51.

37. *London Observer,* 25 May 1969.

38. Deborah Jowitt, in *The Village Voice* on October 20, 1975.

39. *New York Times,* 19 April 1956.

40. *New York Times,* 30 April 1955.

41. Caroline Bristol Britting in a letter to the author, January 15, 1991.

42. Mary Farkas, *Choreography and Dance* (1989), 61.

43. *New York Times,* 23 March 1959.

44. *New York Herald Tribune,* 23 March 1959.

45. *Dance Perspectives,* no. 18, p. 67.

46. Rambert, *Quicksilver,* 200.

47. Joel Kasow, "American Ballet Theatre: A Chronology, 1959–1964," *Dance Chronicle,* vol. 2, no. 4 (1979), 281.

48. *Dance Perspectives,* no. 18, p. 67.

49. Antony Tudor in interview with Marilyn Hunt, August 1986.

50. *Dance Perspectives,* no. 18, p. 65.

51. Arthur Todd, *Dance Observer* (January 1961), 8.

52. Arthur Todd, "Ballet at the Met," March 1961, *Dance Observer,* 43.

53. Tudor, "Movement in Opera," *Opera News* (11 February 1961), 10.

54. Selma Jeanne Cohen, *Dance Perspectives,* no. 18, p. 68.

55. Alwin Nikolais, (1912–1993) contemporary choreographer, who composed electronic scores for his dances.

Chapter 8, Tudor Extends Himself

1. Mary Farkas, interview with author, January 1988, New York City.

2. A year earlier, he had been back in Stockholm briefly to restage *Dark Elegies,* 6 September 1961.

3. Anna Grete Stahle, quoted in *Dance Perspectives,* no. 18 (1963), 86.

4. The author began rehearsals in the role of Serafina wearing the white mantilla; Tudor was at his best rehearsing for this dance—full of clever and delightful parries, good-humored and energetic.

5. Tudor's notes from the Notation Score.

6. Mary Clarke, *The Dancing Times* (September 1963), 4.

7. Jane Pritchard, letter to author, January 1991.

8. The ballet will be called *Echoing of Trumpets* in our discussion.

9. Selma Jeanne Cohen, "Time for Dance in Stockholm," *Saturday Review of Literature* (26 June 1965), 55.

10. Jack Anderson, *Dance Magazine* (May 1966), 42.

11. Ibid.

12. See Aanya Adler Friess, "Echoing of Trumpets" unpublished paper, University of New Mexico, 1988.

13. Anna Greta Stahle, quoted in *Dance Perspectives,* no. 18 (1963), 86.

14. *New York Times,* 10 November 1964.

15. Jack Anderson, *Dance Magazine* (May 1966), 42.

16. Tudor writes in a letter of October 1963 (in the Dance Collection) at the NYPL for the Performing Arts that the Germans "do not know the Classics—their past is built on the Tatiana Gsovsky repertory and style."

17. Edward Villella's story about his unpleasant encounter with Tudor during rehearsals for *Dim Lustre—Prodigal Son: Dancing for Balanchine in a World of Pain and Magic* (New York: Simon & Schuster, 1992), 157–161.

18. *New York Times,* 7 May 1964.

19. *New York Times,* 24 May 1964.

20. *Dance Magazine* (May 1966), 11.

21. Sally Brayley Bliss, "Personal Reminiscences," *Choreography and Dance,* London (1989), vol. 1, part 2, p. 31.

22. Ibid.

23. Ibid., 32.

24. Remindful of his Juilliard piece "Passamezzi" (Gradus ad Parnassum, 1961).

25. *New York Times,* 28 March 1966.

26. A. V. Coton, *Writings,* 73.

27. Selma Jeanne Cohen, "Tudor and the Royal Ballet," *Saturday Review of Literature* (13 May 1967), 74.

28. "Shadowplay," *The Village Voice,* 18 May 1967, p. 18.

29. *Soho Weekly News,* 3 May 1979.

30. They may be found in Dance Collection at the New York Public Library for the Performing Arts.

31. Ibid.

32. Cohen, "Tudor and the Royal Ballet," 75.

33. Jack Anderson *Choreography Observed* (Iowa City: University of Iowa Press, 1987), 102.

34. Mary Clarke, *Dancing Times* (March 1967), 290.

35. Peter Williams, *Dance and Dancers* (March 1967), 12.

36. Clarke, *Dancing Times,* 291.

37. Williams, *Dance and Dancers,* 12.

38. Remy Charlip, *The Village Voice,* 18 May 1967, p. 18.

39. P. Williams, *Dance and Dancers,* 12.

40. *London Times,* 26 January 1967.

41. [London] *Sunday Times,* 29 January 1967.

42. A. V. Coton, *Writings on Dance,* 74.

43. *New York Times,* 3 May 1967.

44. Croce, *New Yorker Magazine* (11 August 1975), 70.

45. Yet we know that Tudor had been shockingly sexual in many of his earlier ballets—*Gallant Assembly, Pillar, Undertow*—to say nothing of the treatment of women in *Echoing* or the whores in *Judgment of Paris.*

46. John Percival spoke with Tudor about Ashton inviting Tudor to make a ballet for the second company in *Dance and Dancers* (March 1969), 28.

47. Peter Williams, "On the Royal Road," *Dance and Dancers* (March 1969) 26.

48. A. V. Coton, *Writings on Dance,* 74.

49. [*London*] *Times,* 2 March 1969.

50. Peter Brinson, *Ballet for All* (London: Pan Books, Ltd.), 1970, 147.

51. [*London*] *Sunday Times,* 2 February 1969.

52. *Dance and Dancers* (July 1982), 26.

53. Maya Deren, *The Divine Horsemen,* Foreword by Joseph Campbell (New York: Delta, reprint 1972).

54. *The Melbourne Age,* 30 November 1969.

55. Edward Pask, *Ballet in Australia: The Second Act, 1940–1980* (Melbourne: Oxford University Press, 1982), 140.

56. Arlene Croce, *AfterImages* (New York: Vintage Books, 1979), 281.

57. Letter to Mirrow Brown may be found in the Dance Collection at the New York Public Library for the Performing Arts. These remarks highlight Tudor's penchant for clarity and details: the color and materials of the women's dresses, their exact lengths, and the atmosphere reminiscent of Chekov's afternoon dialogues. It also reminds the dance historian, as well as the choreographer resetting the piece, that there is no substitute for the words of the choreographer, his thoughts, his direction. Tudor tells us about himself as well as *Sunflowers* and his interest in the sad truth, the patient telling of the social story of members of a group.

58. Tudor, interview by Marilyn Hunt, 22 August 1986.

59. Ibid.

60. Camille Hardy, *Dance Magazine* (September 1986), 86.

61. Ibid.

62. Judith Hansen, newspaper review, Philadelphia, October 28, 1972, Dance Notation Bureau, N.Y.

63. Robert Croan, *Post Gazette,* 28 October 1972.

64. James Felton, *Evening Bulletin,* 17 November 1972.

65. *Dance Magazine* (February 1991), 108.

66. *New York Times,* 24 October 1979.

67. Tudor noted the likeness of *Sunflowers* to his later ballet *Leaves Are Fading,* especially as he chose the same composer, Dvorak. However, *Continuo,* with its emphasis on couples dancing and lyrical classical ballet, seems more the harbinger of *Leaves. Continuo* requires that the dancers learn how to partner each other with a lyrical sensibility. But the dancers must work for clear musical and technical understanding rather than a mature or enduring dramatic statement that *Leaves* demands.

68. *New York Times,* 13 October 1980.

69. Croce, *AfterImages,* 308.

70. Ibid.

71. *Dance Magazine* (August 1987), 18.

Chapter 9. Honors at Last

1. Tudor lectures at the Henry Street Playhouse, April 8, 1951.

2. Carina Ari, who died in 1970, was one of the greatest Swedish dancers. When she died, she left a huge fortune, now the Carina Ari Foundation, which helps the dance and dancers of Sweden.

3. *Dance Magazine* (May 1974), 42.

4. In an article by John Gruen in the *New York Times* on July 13, 1975, Tudor tells us that "it's incredibly hard to make those pieces fit together because of the changes in tonality from key to key. Dvorak composed in too many of the wrong keys."

5. Antony Tudor in interview with Marilyn Hunt, 22 August 1986, New York City.

6. Mary Farkas, *First Zen Institute of America Notes,* vol. XXXIV, no. 5 (May 1987).

7. Gelsey Kirkland, *Dancing on My Grave* (New York: Jove Books, 1986), 198.

8. Ibid., 259.

9. *New Yorker Magazine,* 11 August 1975, 70.

10. *New York Times,* 13 July 1975.

11. "The Heart of the Matter," *Soho Weekly News,* 3 May 1979.

12. Fernau Hall in interview with author, 13 April 1988, London.

13. *Soho Weekly News,* 3 May 1979.

14. *Village Voice,* 21 May 1979.

15. *New York Times,* 13 May 1979.

16. Ibid.

17. Jowitt, *The Village Voice,* 21 May 1979.

18. *New York Times,* 13 May 1979.

19. Dorothy Thom, 17 May 1979.

20. *New York Times,* 10 June 1979.

21. *Ballet Review* (Spring 1983), 63.

22. Letter may be found in the Rambert Archives.

23. Hugh survived Tudor by one year; Nora Kaye died a short time before Tudor.

24. Mikhail Baryshnikov, eulogy, June 9, 1987, Juilliard Theatre Memorial service for Antony Tudor.

Chapter 10. Epilogue: The Heart of the Matter

1. P. W. Manchester, *Ballet Review,* New York (Spring 1992), 74–77.

2. We owe thanks to the Dance Notation Bureau for their efforts at preserving his ballets.

3. Clive Barnes, "The Seal of Greatness," *New York Times,* 23 January 1966.

4. Harold Schonberg, "Balanchine: Slaughtering a Sacred Cow," *Harper's Magazine* (April 1972), 101.

5. Douglas Turnbaugh, *New York Magazine* (20 May 1968), 51.

6. Moira Roth, "The Aesthetic of Indifference," *Artforum 16* (November 1977), 46–53.

7. "Balancing Act," *Dance Magazine* (October 1990), 57.

8. Carla Fracci, Leslie Brown, Martine Van Hamel, Michael Owen, Robert Hill, Johan Renvall, and Kathleen Moore are notable exceptions.

9. Letter from Donald Mahler to author, April 1991. Mahler is a choreographer, teacher, and former Tudor dancer with the Metropolitan Opera Ballet who, in the past ten years, has been helping ballet companies when they remount Tudor's works by coaching them in the Tudor style of dancing.

10. Zita Allen, "The Heart of the Matter," *Soho Weekly News,* 3 May 1979.

Index

Analysis and discussion of Tudor ballets appears in **boldface** numbers, illustrations in *italic*.

Index